Refrigeration Nation

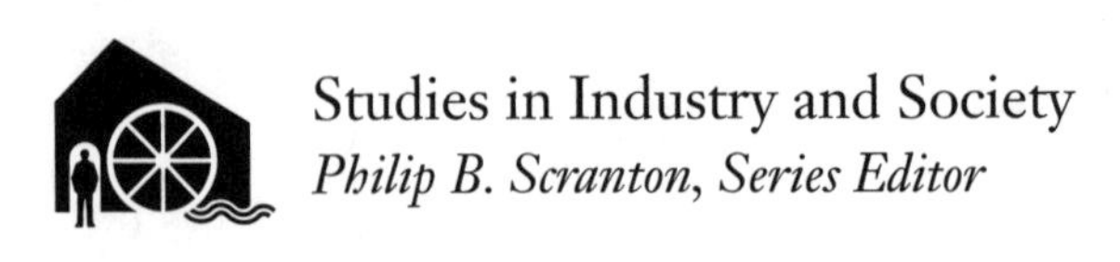

Studies in Industry and Society
Philip B. Scranton, Series Editor

Published with the assistance
of the Hagley Museum and Library

Refrigeration Nation

A History of Ice, Appliances, and Enterprise in America

Jonathan Rees

The Johns Hopkins University Press
Baltimore

Printed in the United States of America on acid-free paper
9 8 7 6 5 4 3 2 1

The Johns Hopkins University Press
2715 North Charles Street
Baltimore, Maryland 21218-4363
www.press.jhu.edu

Library of Congress Cataloging-in-Publication Data

Rees, Jonathan, 1966–
Refrigeration nation : a history of ice, appliances, and enterprise in America / Jonathan Rees.
pages cm. — (Studies in industry and society)
Includes bibliographical references and index.
ISBN-13: 978-1-4214-1106-4 (hardcover : acid-free paper)
ISBN-10: 1-4214-1106-7 (hardcover : acid-free paper)
ISBN-13: 978-1-4214-1107-1 (electronic)
ISBN-10: 1-4214-1107-5 (electronic)
1. Refrigeration and refrigerating machinery—United States—History. 2. Refrigeration and refrigerating machinery—Social aspects—United States—History. 3. Refrigeration and refrigerating machinery—Economic aspects—United States—History. 4. Cold storage industry—United States—History. 5. Cold storage industry—Social aspects—United States—History. 6. Cold storage industry—Economic aspects—United States—History. I. Title.
TP492.7.R435 2014
621.5′640973—dc23 2013006073

A catalog record for this book is available from the British Library.

Special discounts are available for bulk purchases of this book. For more information, please contact Special Sales at 410-516-6936 or specialsales@press.jhu.edu.

The Johns Hopkins University Press uses environmentally friendly book materials, including recycled text paper that is composed of at least 30 percent post-consumer waste, whenever possible.

Contents

Acknowledgments

I have racked up so many professional and personal debts over the course of working on this project that I'm terrified that I'm going to forget someone important, but here at least are the ones I still remember:

All historians doing original research should always start their acknowledgments by thanking the archivists and librarians they have met because where would they be without them? William Worthington from the Smithsonian's National Museum of American History was the person who first told me about the fire at the World's Columbian Exposition while I was visiting there on a travel grant. I am enormously grateful for the assistance of Jan Longone, the Curator of American Culinary History at the Clements Library at the University of Michigan in Ann Arbor, especially in helping me find the materials I needed to finish chapters 7 and 8. I remain amazed that anybody thought to save the instruction booklets that came with old refrigerators and am very grateful that Jan did. Emily Sigman of the American Society of Heating, Refrigerating and Air-Conditioning Engineers helped me long after my only trip to that group's tremendous library in Atlanta. Here at home, Sandy Hudock and Kenny McKenzie at Colorado State University–Pueblo are both interlibrary loan superstars.

Of all the places I worked on this book, I owe the greatest thanks to the Hagley Museum and Library for two separate travel grants and for letting me present sections of the book in workshops there twice. Thanks to Roger Horowitz, Phil Scranton and everyone else at Hagley upon whom I came to depend over the years for help of all kinds.

Hagley also included me in its 2006 Food Chains conference, an experience that proved to be absolutely vital in determining the way I organized the considerable amount of material that I had accumulated by that time. Because of my participation in that conference, the University of Pennsylvania Press published an early version of chapter 3. Similarly, a very early section of chapter 2 appeared in *Technology and Culture*. Other material from chapter 3 originally appeared in *Business and Economic History On-Line*.

Three people involved in refrigerating engineering made invaluable contributions to this book: Andy Pearson, Bernard Nagengast, and Doug Reindl. Without their help I would have made lots of mistakes describing the workings of many different refrigeration technologies. Any mistakes that remain are, as the saying goes, entirely my own.

I think of Roger Horowitz (in his scholarly capacity) and Gabriella Petrick as my best food history friends. They not only fed me sources and helped shape the final form of the manuscript; they read draft chapters when they had many better things to do. For that I am particularly grateful. Scott Martelle also read some of this work in draft for me, offering some of the best writing advice that I have ever gotten. Other historians who have assisted me in my research and writing over the years include Ed O'Donnell, Matthew Morse Booker, Sean Adams, Richard Orsi, Evan Roberts, Suzanne Junod, and Pam Laird.

I participated in panels at three consecutive annual meetings of the Society for the History of Technology during a crucial juncture of my work on this book. In each instance, I took home valuable insights from my fellow panelists and audiences. The "Tasting Histories" symposium at the Robert Mondavi Institute at the University of California–Davis in 2009 proved particularly important to this project, as it was there that I decided that my perspective had to be at least in part global. On a related note, Greg Patmore and the Department of Work and Organisational Studies at the University of Sydney arranged for funding to help me do research on the Australian refrigeration industry. I cannot imagine how I ever would have learned so much about Australia otherwise.

Thanks also to Katy Handley and Jane Townsend for the repeated and continued use of their excellent lodgings in Washington, D.C., and to Tad Baker, who knows the state of Maine better than most people know the back of their hand. Over the years spent on this study, the Office of the Provost at Colorado State University–Pueblo awarded me numerous faculty development grants, which I used to help pay for research travel.

The Johns Hopkins University Press supported me in this work long before it was ready for publication. I especially thank Bob Brugger for his assistance, patience, and excellent editorial advice.

Last, but hardly least, thanks go to my family—especially my wife, Laura—for their love and encouragement. This book is dedicated to my son, Everett, who was not born when I started work on it and whose rapid growth continually served as a reminder that I had been laboring on this project for far too long.

Refrigeration Nation

INTRODUCTION

The Cold Chain

"The producing of cold is a thing very worthy the inquisition," wrote the scientist and philosopher Francis Bacon in 1624. "For heat and cold are nature's two hand's by which we chiefly worketh; and heat we have in readiness; but for cold we must stay 'til it cometh."[1] In Bacon's time, people could produce heat by lighting a fire but could enjoy the cold only in winter (assuming they lived in a place with cold winters). That explains why mastering coldness had been a dream of mankind for centuries before people finally began to do so during the early 1800s. In 1626, Bacon himself tried an impromptu roadside experiment in which he stuffed a dead hen with snow to see if the cold would preserve the carcass. The fact that such a learned man considered this subject worthy of inquiry suggests the rarity of this kind of preservation at that time. The experiment "succeeded excellently well," he reported, before a chill brought on by exposure to that cold led to his death.[2]

Technically, cold does not exist: it is simply the absence of heat. Human beings can produce heat much more easily than they can remove it. As early as the seventeenth century, people understood the potential of the mechanical production of cold for both food preservation and comfort, but creating and controlling cold required literally hundreds of years of scientific experiments. Developing systems to provide reliable transcontinental transportation of perishable products required more experiments. Even mastering cold-related technology constituted only one aspect of the system that made it possible to transport ice and perishable food commodities across great distances. Creating markets for perishable goods preserved by refrigeration required extraordinary risks by entrepreneurs worldwide. Some of these efforts proved unprofitable, but enough succeeded to make refrigeration and goods preserved by refrigeration a vital part of modern life.

Refrigeration has become so prevalent today because it has many advantages over other methods of food preservation. Some of these alternatives existed long before refrigeration did. Drying food probably predates the development of cooking. Salting or curing meat also began in prehistoric times. Cooks developed potting during the 1600s by precooking meat in an airless environment under a skin of fat or butter.[3] These older preservation methods often led to very enjoyable culinary results. For example, making jam or preserves developed as a way to keep fruit from spoiling.[4] The advantages of preserving apples in the form of hard cider should require no explanation.

Despite these happy outcomes, earlier methods of food preservation had many drawbacks. Salting, drying, or smoking destroys many of the nutrients in foods.[5] The flavors that these methods produce cannot pass for fresh. Canning, especially commercial canning, often requires additives like salt or sugar. Early chemical preservatives like sulfuric acid, borax, or even formaldehyde could do serious damage to health when ingested over the long term. Eating badly canned food can lead to botulism and prove fatal. Pack vegetables the right way, in clean sand and hay in a root cellar, and they still could spoil after a particularly hard freeze. Get lucky with the weather, and that cellar kept your root vegetables or seed corn reasonably well, but it did nothing for meat or milk.

Refrigeration through cold temperatures is the only form of food preservation that will not change the food's taste.[6] Refrigeration, mechanical or otherwise, also far surpassed these other methods of preservation in terms of convenience. Pickling, potting, or drying was an enormous amount of work—always for the person doing the preserving, sometimes for the person doing the consuming too. Refrigeration provided the possibility of eliminating that work entirely. Because of the disadvantages inherent in these earlier methods, consumers immediately recognized the value of reliable refrigeration. Had mechanical refrigeration never appeared, earlier methods of preservation would still help keep food from spoiling, but mechanical refrigeration offered the possibility of keeping food fresh longer, more efficiently and more effectively. For these reasons, the perfection of refrigeration constitutes a major milestone in world culinary history.

In its early years, refrigeration seemed almost like magic. It could not extend the life of perishable food indefinitely, but it quickly became the best possible way to preserve such products as the effectiveness of this technology improved. First ice, and later, mechanical refrigeration, raised the prospect of a wider variety of goods appearing on tables worldwide at all times of the year. By cutting waste, ice and refrigeration made more food available without the necessity of devoting more land to food production. As was the case with earlier food preservation technologies, people came to depend upon refrigeration every day, often at every meal. Perhaps they had no icebox or refrigerator in their home, but over time more and more people worldwide took advantage of supply chains that depended upon refrigeration for a substantial part of their diets. Refrigeration became part of the infrastructure upon which everyday life depended.

Refrigerating engineers refer to the infrastructure that makes paths of trade for perishable food possible as cold chains. Cold chains consist of the linked refrigeration technologies needed to preserve and transport perishable food from its point of production to its point of consumption.[7] Sometimes identical technologies refrigerated different things at different stages of the chain. The same ammonia compression refrigeration equipment, for example, worked for anyone who wanted to enter the ice-making business or for someone who wanted to open a cold storage warehouse. The cold came from the same equipment in each case, but buyers adapted the machine in order to fit either of these purposes. Sometimes different technologies refrigerated similar things in different environments along the cold chain. An icehouse near a pond in New England, in which natural ice sat from harvest time in winter until the height of summer, had a different design than an icehouse in the tropics that preserved the same product because the latter structure needed much more insulation. Despite different conditions at different stages along the chain, every business that constituted a link on that chain had an interest in making sure it could preserve a quality product. Otherwise, national and international markets for perishable goods would not have developed.

A cold chain is a kind of food chain, and food chains have much longer histories than cold chains do. To a biologist, a food chain follows energy in the environment in many different ecological forms. Think plants at the bottom and predators at the top. When man developed industrialized food chains, geography became just as important as the ecological processes that produce the energy in the first place. Think of the transport of spices across continents in ancient times. The study of food chains allows for the examination not only of the cultural significance of particular commodities, but also of the scientific, technological, and even political factors that make a particular food available for consumption. The history of food chains includes the history of agriculture, environmental history, economic history, and the cultural history of the people doing the eating.[8]

The dynamics of both food chains and cold chains help explain the history of the American ice and refrigeration industries. Different businesses using different technologies gradually organized long chains to take both ice and foodstuffs on long journeys across America and across the world in order to make a profit. Today, the possibility of consuming food from around the world has faded into the background of everyday life. The electric household refrigerator now serves as the most visible manifestation of the modern cold chain but that is just its end point. Storage, transport, and even the display cases in grocery stores require reliable refrigeration for the cold chain to work effectively. Developing all these links took time and an enormous amount of technological innovation.

The development of cold chains depended upon the development of technology to control the cold. The American ice industry of the early nineteenth century offered the first opportunity in human history for man to harness cold for entrepreneurial purposes. Ice shipments predate food packed in ice, thereby making ice itself the first highly perishable commodity with a world market. Ice constituted part of a cold chain by definition since it required protection (provided by more ice) to survive an ocean journey of any significant distance. Other food products that depended upon ice to remain cold sometimes got transported without refrigeration, but their incorporation into cold chains made their shipment more effective and profitable. Entrepreneurs could make full use of a cold chain only after the technologies used all along it became effective and when such chains had expanded worldwide. That process took less than two hundred years to complete.

The Four Stages of Cold Chain Creation

Even moving perishable food a short distance from the point of its production to the point of its consumption required overcoming many obstacles. The extent of those obstacles explains why technologies along the cold chain developed at different rates. In most cases, these changes occurred very gradually. As a result of this slow pace of change, the cold chain had large gaps, with no possibility for refrigeration for many perishable foods until as late as the 1920s. When any product reached the table in a less-than-ideal state, it did not really matter where the problem occurred. Consumers often blamed refrigeration technology, and this called the value of the entire cold chain into question. To avoid this situation, everyone involved in making the cold chain function—food producers, transporters, storage companies, equipment manufacturers—had every incentive to work together. When refrigeration served the interest of consumers, it became profitable to all of them.

Despite this incentive, many refrigeration-related companies failed to make money before the various links along the chain developed into their modern forms. What seemed like a good idea in theory to Francis Bacon in 1624 did not always prove profitable in practice. Sometimes consumers got too impatient while waiting for the perfection of a particular technology. Sometimes entrepreneurs could not create markets for new products that most consumers did not desire or could not afford. Sometimes the demand for a particular good got too far ahead of the technology that existed to preserve and transport it. These problems slowed down the pace of technological change even as the demand for refrigeration in general grew stronger over time.

Notwithstanding such setbacks, the modern cold chain emerged between 1806 and the present. This long, convoluted process went through four separate, overlapping stages. The process began when people learned to generate cold for the first time. Surrounding something with ice makes it colder. Ice therefore became the first source of industrial cold and the first product to travel along a cold chain. In places where the lakes, ponds, and rivers never froze, the cold chain often made ice available for the first time. It still seems surprising that ice, packed well, could travel from New England all the way to India intact, yet ice harvesters shipped their product along this route regularly starting in the 1830s. Once scientists could create artificial cold, the effective preservation and transport of perishable foodstuffs became possible too. It took time, however, before mechanical refrigeration equipment became affordable and effective. This allowed the ice industry to survive well into the age of mechanical refrigeration.

Once efforts to create cold proved successful, people began to learn to manage the cold too—the second stage of cold chain creation. That meant delivering cold to a particular space for an extended period of time. If perishable food spoils once, it remains spoiled forever. That explains why, after the creation of artificial refrigeration, making that technology reliable became the refrigerating engineer's top priority. Refrigerating engineers also worked hard to shrink refrigeration machines in order to make them both more affordable and more convenient. The first marketable refrigerating machines weighed tons. These monsters usually produced ice for other parts of the cold chain like railway cars rather than for individual consumption. Mechanical refrigeration formed the infrastructure of the cold chain before refrigerators ever became a consumer product. Developing mechanical refrigeration for ships, for example, made the transcontinental trade in all perishable products—especially meat—possible.

Shrinking refrigerating machinery for shipboard purposes or other uses took decades. Developing mobile land-based mechanical refrigeration took until well into the twentieth century. A reliable mechanical household refrigerator took almost as long because engineers had to reconceive the entire technology to generate cold. After more compact refrigeration became available, however, consumers benefited from the considerable convenience that household refrigerators could provide. Had it cost too much to refrigerate perishable products at any stage along the cold chain or had the equipment at any of these stages remained unreliable, the development of the cold chain would have ground to a halt.

As the ability to manage the cold became possible, people learned to control the temperature with precision. This represented the third stage of cold chain creation. Different foods keep best at different temperatures. Keep them all at the same temperature, and some of them will taste less than ideal as a result. Icebox owners faced this problem during the mid-nineteenth to early twentieth centuries, since no way existed to control the temperature of a block of ice. Cold storage facilities, way stations along the cold chain, dealt with this same problem as the volume and variety of goods transported along cold chains expanded after 1900. Freezers, devices that keep perishable goods particularly cold, also needed precision in temperature. Only the right amount of cold delivered to the right place could ensure that the perishable food shipped along the cold chain remained as fresh as possible under the circumstances. The greater the variety of goods shipped along the cold chain, the more important it became to control the cold with precision, as increased volume made it harder to refrigerate every part of any shipment to its ideal temperature. Cross-contamination, of the sort that might happen if you stored fish and eggs together, proved equally problematic because foods at the same temperature can spoil at different rates. Once these problems got solved, the use of refrigeration expanded considerably.

With these goals achieved, the many businesses responsible for the cold chain began to expand and extend its volume and reach. Both these kinds of growth represent the final stage in the creation of the cold chain and continue even today. Referring to a single cold chain is a linguistic convenience. In reality, the ice industry, in its earliest forays into preservation through cold, and nowadays the refrigeration industry built many cold chains connecting producers and consumers around the world. In their early years, these chains sometimes contracted in order to improve their reliability. In recent years, they have almost always expanded. Today, any perishable good can travel from one part of the world to another with the only damage inflicted upon it coming from the inevitable effects of low-temperature preservation upon taste. Even though businesses cannot make money shipping every perishable product, they have profited by doing so for many of them. The result has been the expansion of cold chains around the world in both length and number. Every major agricultural production hub in the world currently serves as the starting point for cold chains for products of all kinds. Just go to your local supermarket and read the country-of-origin labels on the packaging or the signage. The huge variety of foods available in countries at the ends of these cold chains serves as one of the most potent symbols of modern globalization.

At all of these stages, the history of the cold chain in America and the world has depended upon the interaction between technology, business, and consumers. This book follows different clusters of refrigeration-related technologies in each chapter, sometimes considering a single grouping, sometimes considering different kinds of technologies that competed against each other in the global marketplace. Each chapter examines the advantages and disadvantages of different segments of the cold chain and describes the effects that these technological changes had on the consumption of perishable foods in other countries around the world. The book begins by explaining how nineteenth-century Americans generated cold, first with ice, then by using mechanical refrigeration. Then it moves down the cold chain, explaining how perishable goods got transported and stored. Finally, it covers the end point of the cold chain, explaining how consumers stored perishable foods in their kitchens, from iceboxes during the mid-nineteenth century to the huge electric household refrigerators of today. The end of the book considers the place of refrigeration in modern capitalism, and the relationship between refrigeration technology and the natural world.

Technological changes have created or accentuated the effects of the cold chain, both good and bad, upon societies everywhere. Even though refrigeration is now available worldwide, the United States has experienced all of these effects to a greater extent than any other country in the world. To say that the United States is a "refrigeration nation" does not mean that no other country played a role in the development of the modern cold chain or that America is the only nation that benefits from refrigeration technologies. What it does mean is that the United States deserves most of the credit for the development of the modern cold chain and that it has made the most use of refrigeration technology over time. The earliest cold chains tended to run into, out of, or through the United States. The evolution of this technology in the United States determined what technologies became available for use elsewhere, even if the adoption of those technologies on a worldwide scale depended largely on local factors. Nor have all Americans seen refrigeration technology the same way. Refrigeration changed everyday life wherever it took hold, but what you thought about such changes often depended upon what form that disruption took. Social class, gender, and especially culture have played a huge role in how people perceived such changes.

While this book concentrates upon the history of ice and refrigeration in America, people from many other countries played an important role in the development of the cold chain too. For this reason, this work includes sections on ice and refrigeration around the world, mostly as reported by American sources. The ways Americans made use of refrigeration technologies spread elsewhere, albeit more slowly and in some cases not to the same degree. Brief sketches of the ice and refrigeration industries in countries outside the United States also serve as counterpoints to demonstrate the importance of refrigeration to American industry and culture. While the United States took to refrigeration of all kinds faster and more extensively than the rest of the world, the tremendous advantages of refrigeration have made it increasingly popular in developed countries around the globe in recent years. What follows, then, is not a global history of ice and refrigeration, but more a history of the American ice and refrigeration industries in a global context.

From the Foreground to the Background

The cold chain changed what America and the world ate forever. Despite its obvious importance, the American public became oblivious to the extraordinary effects of refrigeration technology very quickly. As the *Milwaukee Journal* remarked in 1894, "What is really a modern wonder has become such a common thing in business, like the telephone, the public hardly notices it."[9] Over the course of the nineteenth century, ice went from a luxury to a necessity for the vast majority of Americans. During the twentieth century, refrigeration became still more important even while blending further into the background of day-to-day existence. Examining the ice and refrigeration industries from a historical perspective allows us to look at this vital technology with fresh eyes. Even when new refrigeration technologies did not work very well, people accepted drawbacks associated with refrigeration that modern consumers would find maddening. Consumers began to take these technologies for granted as the industry worked out the kinks. By recapturing the initial, fleeting sense of wonder people felt when they first saw how refrigeration could change their lives, we can consider the role that refrigeration plays in today's world with the same appreciation for its effects that only its very first beneficiaries saw clearly.

Thanks to refrigeration, consumers added tremendous variety to the previously monotonous American diet. Walter Sanders's work for the trade journal *Ice and Refrigeration* made him more contemplative about the effects of ice and refrigeration on American society than most of his countrymen. His 1922 argument that "ice is responsible for the ringing of the death-knell of the old menu" still captures the amazement that observant Americans must have felt if they recognized how what they ate had changed over the course of a single lifetime.[10] The economist Simon Patten took particular note of how refrigeration offered greater food choices even to the poor: "In the Philadelphia ghetto one may watch a huckster's wagon, filled with strawberries costing three and four cents a box, move down the block and go away empty on a day when a box uptown is not be found at less than ten cents. Other wagons offer lettuce, celery, corn and spinach, banked crisply against the blocks of ice in the wagon bed. Banana carts make their rounds during the whole year with excellent fruit for ten and eight—and late on market nights for six—cents a dozen."[11] The more effective the cold chain became, the more prices dropped, thereby making its benefits more widespread; and the variety of food that was made possible greatly improved the nutritional intake of people of every social class.

As inventors, entrepreneurs, and businesses made the cold chain more reliable and efficient, American consumers reacted with both approval and resignation. Sometimes they supported technological changes as they occurred because they saw how these changes would improve their lives. Occasional resistance—either because people did not recognize the advantages of the new technology at first or because the disadvantages seemed too great—never lasted long. The magnitude of potential benefits from new refrigeration technologies has determined the extent and the speed of the transitions from one technology to another both in America and elsewhere. The more advantages a technology offered, the more likely consumers of refrigeration wanted to adopt it quickly. If it took time for a new technology to operate to its full potential, the transition occurred much more slowly. Because this book focuses on such periods of technological transition, most of the story falls between the 1870s and the 1920s, the period when change in the ice and refrigeration industries came at the greatest speed. During this era, the technologies that constitute almost every part of the modern cold chain possible debuted in one form or another.

The modern cold chain has not developed along a straight line of steady progress but, rather, through a process of technological evolution devoted to making different technologies more effective and more profitable. Even today, if a consumer wants only to preserve a picnic lunch through the morning, simple technologies outside the modern cold chain (like an Igloo cooler) will do. Because earlier food preservation techniques work to some degree, their usage often continued well beyond the introduction of their competitors, particularly in light of initial problems with some new forms of refrigeration. Eventually, however, most consumers willingly paid for added value. Technological changes in the refrigeration industry occurred much more slowly outside the United States, since consumers elsewhere in the developed world often cared more about the taste of what they ate rather than speed or convenience. In contrast, Americans' devotion to refrigeration helped define them as American. In other words, the importance of refrigeration for understanding American history goes well beyond its effects upon diet.

Despite its importance, this subject has received little attention from historians in recent decades.[12] One historian who has addressed the significance of refrigeration suggests that its importance on a global scale may match that of the American Civil War.[13] While this book does not make that argument, it does describe how the ice industry kept us cool in the heat before air conditioning; how refrigerated transport let people eat perishable foods grown far away; and how cold storage helped us defy the seasons and place a greater variety of foods on the table throughout the year. These achievements changed countless people's lives for the better. In the course of describing such benefits, this book also discusses some of the costs associated with the growth of the modern cold chain. Whether the benefits of refrigeration justify America's long, fervent embrace of this technology depends upon your priorities. Nonetheless, understanding how the cold chain developed should prevent you from ever taking the effects of refrigeration for granted again.

CHAPTER 1

Inventing the Cold Chain

On a cold winter morning in the late 1840s, one hundred Irish immigrant laborers and their native-born American foremen arrived at Walden Pond in northeastern Massachusetts, where Henry David Thoreau lived in a small cabin that he had built himself. The men, Thoreau noted, came "with many car-loads of ungainly looking farm tools, sleds, ploughs, drill-barrows, turf-knives, spades, saws, rakes, and each man was armed with a double-pointed pike-staff. . . . I did not know whether they came to sow a crop of winter rye, or some kind of grain recently introduced from Iceland."[1] Thoreau knew they were not farmers, at least not in the strictest sense of that word. They were ice harvesters who had come to carve out frozen blocks of water from the pond to stack in icehouses on the banks, where the blocks would sit until the weather warmed. The writer who went to live in the wilderness to get away from a newly industrializing society had to deal with a new industry which had literally come to him.

The workers who disturbed Thoreau's isolation at Walden Pond cut ice for sixteen days, working from sunrise to sundown. "They divided it into cakes by methods too well known to require description," Thoreau wrote, "and these, being sledded to the shore, were rapidly hauled off to an ice platform, and raised by grappling irons and block and tackle, worked by horses, on to a stack, as surely as so many barrels of flour, and these placed evenly side by side, and row upon row, as if they formed the base of an obelisk designed to pierce the clouds. They told me that in a good day they could get out a thousand tons, which was the yield of about an acre."[2] The methods of the natural ice industry were indeed "too well known to require description" when Thoreau published *Walden* in 1854, but most people have forgotten them now.

An ice industry arose in New England between the turn of the nineteenth century and the time of Thoreau's stay on the shores of Walden Pond. This activity began the development of the modern cold chain because for the first time people cut ice to manage the cold for commercial purposes. While the early natural ice industry achieved little precision or efficiency, it created remarkably long cold chains, even by modern standards. The melted parts of the ice shipment protected the rest of the ice packed in with it. The geographic reach of the ice industry that harvested its product outside his door particularly impressed Thoreau. "Thus it appears," he wrote in *Walden*, "that the sweltering inhabitants of Charleston and New Orleans, of Madras and Bombay and Calcutta, drink at my well."[3] If anything, the inhabitants of Charleston, New Orleans, Bombay, Madras, and Calcutta probably marveled at this feat even more than Thoreau did, since they seldom saw ice at all.

By the 1850s, ice harvesters around the world used the same methods as the first ice dealers in New England to harvest, transport, and store their product. First, the harvesters had to remove any snow that had accumulated on top of the ice. Once they scraped that off, the harvest team had to measure the thickness of the ice with an auger. They only harvested ice a foot thick or wider, though the standard could shrink to eight inches during particularly warm winters. (Anything thinner than eight inches would have melted before it ended its travels through the cold chain.) Before cutting, they marked out a grid pattern on the ice to guide the saws.[4]

Once the grid was marked, teams used a single-axle plow strapped to a horse to cut furrows in the ice similar to those plowed on a farm field. Foot-thick ice could hold the weight of several workhorses (shod with spiked horseshoes) and many harvesters without cracking. Although very deep, the furrows cut by the plow did not go all the way through to the bottom of the ice at this stage of the harvest. Workers used ice chips to caulk any holes that the cutting had made visible between the partially cut blocks of ice. This prevented the grooves from filling up with water and freezing shut. Then they used a very long ice saw or a bar chisel to break off the first chunk of ice in a place near the icehouse. In a placid body of water like Walden Pond, they would have removed the first pieces in a linear pattern in order to form a channel. Creating a channel made it easier to float the remaining pieces of ice to the icehouse before removing them from the water. The horses that Thoreau saw would have dragged the blocks of ice out of the water, where they would have been cut into smaller pieces and stacked in icehouses along the shore. There they would have sat until summer. Then, after traveling to Boston by rail and being packed into ships, the ice would have traveled to ports around the world.[5]

Even late in the nineteenth century, ice harvesting techniques still resembled the way in which farmers harvested crops. Horses on the ice pulled ice cutters that resembled plows. From J. T. Trowbridge, *Lawrence's Adventures among the Ice-Cutters, Glass-Makers, Coal-Miners, Iron-Men and Ship-Builders* (Boston: Fields, Osgood & Co., 1871). Courtesy, The Winterthur Library: Printed Book and Periodical Collection.

The man responsible for harvesting ice on Walden Pond during Thoreau's stay along its shores was named Frederic Tudor. Tudor pioneered every aspect of the natural ice business—harvesting, shipping, and marketing ice to other Americans and people throughout the world. Without his groundbreaking efforts to get Americans to purchase ice, the modern cold chain never would have developed because not enough people would have demanded cold things. Unfortunately for Tudor, it took years of trial-and-error experimentation to create a cold chain that could deliver ice effectively. Even in cold climates, availability became a major impediment to consuming ice since only a rudimentary cold chain existed once the product landed in port. Price constituted another impediment to ice consumption; if prices got too high, consumers cut back their purchases. In order to keep ice available and affordable, Tudor had to overcome significant technological difficulties, particularly to ship his cold cargo between continents. Once the market for natural ice developed, the cold chain shortened as more entrepreneurs entered the industry. Americans developed a taste for ice, and the natural ice industry became increasingly profitable.

A "Slippery Speculation"

Frederic Tudor began his efforts at establishing an American natural ice industry with global rather than local ambitions. This made him way ahead of his time. Initially, Tudor's reputation suffered greatly because so few people believed that ice would ever become a necessity, even in America. Yet Tudor imagined a cold chain before any part of that chain even existed. Once he became successful, the natural ice industry he created attracted attention as a lesson in Yankee ingenuity. A region of the United States with few natural resources except its waning supply of timber and the fish in its seas managed to take something that most people had thought valueless and turn it into a major global industry. Every segment of the modern ice and refrigeration industries owes its existence to Tudor's pioneering work towards controlling the cold.

Nobody before Frederic Tudor had ever conceived of a cold chain, which explains why nobody before Tudor had ever tried to build one either. People had cut ice from lakes and stored it in ditches and basements for use during the summer before Tudor came along. But Tudor made cutting, transporting, and selling ice efficient enough to become profitable. His success exporting ice from New England led to the development of local ice markets throughout the country. However, success did not come easily for this refrigeration pioneer. Tudor's career had intense highs and lows over half a century. Once Tudor set out to create an industry where none had existed before, he did not turn back despite financial setbacks and the derision of his peers. His creditors got him imprisoned multiple times before the market for ice became stable, but Tudor went on to become one of Boston's richest men.

Born in 1783, the son of a Boston lawyer who had served on George Washington's staff during the American Revolution, Tudor did not have to pull himself up by his bootstraps to enter the business world.[6] While his grandfather John Tudor, the first Tudor to emigrate to America, had come from humble stock, John left Frederic's father, William, a small fortune of $40,000 when he died in 1794. Frederic's older brother, the second William Tudor, went to Harvard, but Frederic quit school at the age of thirteen and entered the world of business as a clerk at a Boston retail store. Through his father, Frederic eventually got an unpaid position working for a Boston mercantile firm. While working for that company, Tudor used money he got from his family to sponsor voyages that traded in goods like nutmeg, flour, sugar, and tea. Therefore, he was already an experienced merchant before he began trading in ice. Despite his eventual success in the ice trade, the natural ice industry remained speculative even after a market for ice became established. That explains why Tudor never ceased trading in other products. Nevertheless, ice would constitute enough of his capital for him to be universally known as the "Ice King" towards the end of his career.

The reliance of the natural ice industry on the weather made Tudor's business particularly unstable. Since any industry dependent upon the weather will always go through cycles of boom and bust, Tudor documented and studied the weather conditions around Boston very closely every winter. Cold winters made for thicker, more marketable ice and more of it as well. A warm winter, or even just a warm snap during harvest time, could destroy his revenues for an entire year. Tudor and every ice merchant that followed him also had to hope for hot summers everywhere they sold their product in order to make a profit, because higher demand made it possible to charge a higher price. On the other hand, any competitor with a melting stock of ice might severely undercut Tudor's prices out of desperation, since merchants often offered it at a steep discount at the end of the season. Sometimes Tudor undersold his competitors in particular markets in order to drive them out of business so that he could raise prices later.[7] He could do this because the market for ice in the early nineteenth century revolved entirely around the port cities that received shipments of ice from different harvesting concerns. Try to get ice from the next town over during the summer if your supply ran out, and it would likely melt by the time it arrived to fill the demand.

Tudor's first shipment of ice in 1806 to Martinique in the West Indies illustrated all the perils of selling such a perishable product without a cold chain in place. "No joke," wrote a *Boston Gazette* reporter after that first shipment of ice left port: "A vessel with a cargo of 80 tons of ice has cleared out of this port for Martinique. We hope this will not prove a slippery speculation."[8] The first boat that Tudor chartered canceled its contract for fear that the ice would melt and sink the vessel. (That unwarranted fear persisted for decades, even after the industry had substantially grown.) Tudor had wanted the government in Martinique to grant him a monopoly on the ice trade there to allay some of his risks in this unprecedented venture. Instead, he only got permission to sell his cargo directly off the boat. Without an icehouse on the island to slow the rate of melting, the cargo disappeared quickly when the open hold exposed the ice to the warm tropical air. Perhaps he might have sold more if the citizens of Martinique had had a way to stop the ice from melting once they brought it home. On the handbill he used to advertise the sale, Tudor recommended that purchasers carry it back in a blanket or linen cloth to keep it intact, but he still got complaints, since it melted anyway. Because Tudor had permission to sell ice in Martinique for only three days, he had to go to another island to sell what remained in the hold at the end of his stay. In the end, Tudor traded much of the ice for $3,000 worth of sugar to take back to Boston. The total loss for this initial venture was about $2,500.[9]

Despite this shortfall, Tudor immediately began to plan shipments of ice to other markets. Over time, he developed the components of a cold chain to prevent the problems that surfaced at Martinique from vexing him again, but good solutions took decades to develop. Interrupted briefly by the War of 1812, Tudor gradually set up a network of icehouses and icehouse keepers in port cities and islands all over the world. Even in the early stages of the business, Tudor's business took on a radial shape—a series of linear cold chains between Boston and a slowly growing number of warm weather ports. Once Tudor established himself in any market, shipments of New England ice would arrive there throughout the summer. He did not send ice to a new market unless he had placed an agent and an icehouse there first. Havana came first, then Charleston, New Orleans, and Augusta. By the 1830s his ice had proved popular enough that he branched out to sell it in more distant ports. Rio de Janeiro did not get nearly as much ice from Tudor as Calcutta did on the other side of the globe because it lacked the large community of wealthy Englishmen who became Tudor's best customers in India.[10]

While Tudor stayed mostly around Boston—sending correspondence to his icehouse keepers, working to improve the efficiency of his cold chain, and checking the weather—his name became famous around the world. Since he had to arrange every step in the cold chain and introduced ice before anyone else into most markets, he got nearly all the credit for establishing the natural ice industry once it took hold. Grateful English customers in India even sent him a loving cup in his honor. Tudor's competitors sold ice in Manila, Singapore, Canton, Mauritius, and Australia, although few made much money doing so.[11] Tudor's many years of establishing a cold chain finally enabled him to turn a profit selling ice as New England's "Ice King."

Learning through Failure

The few existing biographical sketches of Frederic Tudor tend to emphasize his foresight in seeing a market where none had existed before and his determination in spite of many setbacks, but they underestimate the importance of his technological achievements. Anyone could preserve ice on months-long journeys to tropical climates, since the ice that melted shielded the ice that stayed intact. The more difficult obstacles to overcome came at the beginning and end of these rudimentary chains. Tudor not only had to transport ice out of Boston to points south; he also had to arrange to have it cut, stored by the waterside in Massachusetts until the summer, and then transported to the dock. When summer came, he then had to store his cold cargo at its destination until it sold and, perhaps most importantly for the success of the industry, make sure that a market existed for ice in the places where he sold it.

Tudor had to slow the speed with which his ice melted in order to succeed. His first icehouses in the tropics were built entirely above ground. They were usually made of inexpensive materials such as wood, and the sides were insulated with wood shavings or straw. Because they had to be loaded at the top, these icehouses proved particularly effective at slowing melting because less ice got exposed to the tropical air when it was loaded or unloaded for sale. The opening was reached by stairs that went along the outside of the house. The storehouses were cone-shaped on the bottom "like the roof of a house inverted" to facilitate the draining of the melt water. "Moisture always carries heat," Tudor wrote in 1805. Since melted water was warmer than ice, removing the meltwater meant that the ice could remain solid much longer. Sometimes Tudor even had his icehouses erected inside other buildings so as to further protect his product from the elements.[12]

Apart from storage near the points of production and consumption, Tudor pioneered new ways to store ice better in transit. He first tried packing ice in rice, wheat chaff, and bark before finally settling on sawdust as the best way to keep heat out of his icehouses and the holds of the ships he chartered.[13] He packed vessels only on the coldest days and used enough men to make the process go as fast as possible so that as little ice as possible would melt during transit. The sawdust covered the blocks the same way that mortar separates bricks in a wall.[14] Once the industry became established, Tudor's firm filled ice ships from the bottom of the hold all the way up to the deck. He used hay or small ice shavings cut from larger blocks to make sure that no room existed for warm air to penetrate the cargo hold. This essentially created one huge block of ice made up of individual units that could be easily separated once the ice shipment arrived at the appointed destination. Bills of lading, the contracts under which ship owners carried merchants' cargo, describe what the crew had to do in order to preserve the ice in transit. First, they kept the hold closed throughout the voyage to prevent the effects of warm air on the perishable cargo. Crews also pumped out the meltwater every few hours after packing until the ice got unloaded, not just because the water was warmer than the ice it came from but also because enough water in the hold really could sink a ship.[15] With good weather and swift sailing, an ice ship on a voyage to as far away as India might lose as little as 35 percent of the cargo. With bad luck, that figure could go as high as 75 percent.[16]

Tudor also pioneered a variety of methods designed to harvest more ice faster. The faster the ice was harvested, the less likely it was that warm weather during harvest time would make a once-good ice crop partially melt and thereby become completely unmarketable, since ice pockmarked with air holes could not be cut or stored easily. Pockmarked ice also melted faster. Until the mid-1820s, ice harvesting required lots of labor with handheld tools and took most of the winter. Nathaniel Jarvis Wyeth, a supplier whose parents owned a hotel along Fresh Pond in Cambridge, Massachusetts, and who later became Tudor's employee, significantly improved upon these methods. Most importantly, Wyeth invented a better ice plow that resembled a carpenter's instrument to cut grooves in wood. Wyeth increased the size of the plow and hooked it up to horses for use on the ice. He also created an elevator system to raise large blocks of ice into an icehouse.[17] Over the course of the century, some of the production process became automated to help satisfy the growing demand for ice. Workers still did most of the operations like caulking the sides of blocks with ice chips or dragging the blocks through the exposed water towards the icehouse by hand. Despite some mechanization of the process after the ice left the water, Wyeth's harvesting system remained almost entirely intact throughout the life of the natural ice industry.

Taken in their entirety, Wyeth's innovations led to a huge increase in the efficiency of natural ice production. By 1880, ice companies could cut 100,000 tons of ice in the same time that it had taken to cut 10,000 tons earlier in the century.[18] Wyeth's many new tools made it possible to cut ice into regular shapes with flat sides rather than the irregular product created by ordinary ice picks. Laid cheek by jowl in icehouses or packed into ships, more ice fit into the same spaces. With the cold surfaces in constant contact with one another, all the ice stayed cooler longer.[19]

As a result of this increase in volume, the scale of Tudor's ice-cutting operations gradually increased to the approximately one-hundred-man team who showed up outside Thoreau's cabin. Such laborers, often farmers and farm laborers who had nothing else to do in the dead of winter, worked only on a temporary basis. In areas where the ice did not get particularly thick, harvesters could cut it in as little as six days.[20] In a particularly cold year, companies sometimes harvested ice off the same body of water twice in one season.

Tudor also paid special attention to keeping his ice clean. Writing in *American Notes*, the 1842 account of his first trip to the United States, Charles Dickens described "the great blocks of clean ice which are being carried into shops and bar-rooms" in New York City.[21] That ice almost certainly came from the Hudson, and the fact that such high-quality ice went for commercial purposes along the cold chain suggests just how few households bought ice at that stage of the industry's development. Had Dickens examined the ice close up, it would not have looked quite so clean. Sediment and other natural phenomena such as air bubbles made transparent natural ice hard to find, even in the early nineteenth century. It would be natural for the American public to want transparent ice as a way to assure its purity, yet the desire for transparent ice actually predated concerns over the safety of the product. People did not want sediment in their ice whether it constituted a health threat or not. After all, who wants to see a ring around the glass after finishing a drink? That explains Tudor's excitement in 1830 when he wrote in his diary, "Wyeth tells me the ice is so fine that you can see [and] read a printed paper through a block 42 inches long."[22] Clear ice sold for more money. Usually harvesters could get clear ice only by cutting it from streams, where the current kept both air and sediment out of most of the ice that formed. Most of the natural ice offered commercially came from inferior sites, especially after competition developed during the mid-nineteenth century because only a limited number of good sites for natural ice existed.

Innovations all along the cold chain eventually made this very difficult business profitable. Tudor obsessively managed the entire cold chain that he had created as efficiently as possible in order to keep prices low. His former icehouse keeper in Havana, John W. Damon, became disgruntled for precisely this reason. As Damon later explained, Tudor "had 'all along' endeavored to impress me with the belief that in management"—that is, the management of providing the ice and shipping it to Havana—"lay the secret of all prosperity; that it was by far the most difficult part of our business; one, before the difficulties with which I had to contend were trivial and insignificant."[23] The complete cold chain that Tudor pioneered included not only ocean vessels but also the trip from the lake where the ice got harvested down to the harbor. As early as 1836, during the shipping season ice was hauled from inland lakes to the ports of Boston and Charlestown all day, at half-hour intervals, by six-horse wagon teams.[24] In 1841, a consortium of businessmen led by Nathaniel Wyeth, by then a competitor of Tudor's, extended the Charlestown Branch Railroad out to Fresh Pond.[25] This made it possible for harvesters to significantly increase their volume. But shipping in volume was valuable only if there was a demand for ice in the torrid regions of the planet, so Tudor and his competitors had to pioneer innovative marketing techniques too.

Manufacturing Demand

People around the world had a long history of consuming and using ice before Fredric Tudor created the natural ice industry. In ancient times, many societies cut it from glaciers or froze water in pans if the temperatures dipped below freezing. Underground ice houses existed in China as early as 1100 BCE. Greeks and Romans picked up the ice habit and the design for icehouses from civilizations in the Middle East. Roman emperors had ice delivered to their palaces by donkey from as far away as the Alps. The Turkish Empire shipped boatloads of snow into Istanbul during the sixteenth century, and merchants sold the product in the streets at a price that all classes of consumers could afford. Confectioners made ices for the wealthy merchants in sixteenth-century Florence and at the court of Louis XIV in seventeenth-century France. European travelers noticed that the Persian Empire consumed enormous amounts of ice, selling it by the donkey load to locals for domestic use.[26]

Despite this broad popularity, ice always had its critics too. The ancient Greek physician Hippocrates wondered "why . . . any one [would] run the hazard in the heat of summer of drinking of iced waters," since they are "excessively cold," and might therefore suddenly throw "the body into a different state than it was before, producing thereby many ill effects."[27] Such sentiments existed in every culture that consumed ice. In fact, this kind of prejudice against ice consumption persisted well beyond Tudor's lifetime.[28] Outside the United States, warnings about the use of ice grew even stronger. Herbert Spencer wrote in his autobiography, published in 1904 (a year after his death), that "there can . . . be little doubt that the habit [of consuming ice water] is an injurious one. In the first place, taking an amount of liquid that much exceeding that required for carrying on the bodily functions, is pretty certain to be detrimental; and in the second place, frequently taking this at a temperature so much below blood-heat, is also pretty certain to be detrimental by checking digestion."[29] The supposed results of ingesting cold things included seizures, blindness, paralysis, heart attack, and apoplexy.[30] Because this notion persisted in countries without much of an ice industry long after ice consumption became common in the developed world, Tudor not only had to deliver his product but market it too. The cold chain made it possible for him to sell ice, but Tudor also had to manufacture demand in order to make his investments in a cold chain worthwhile. In order to create that demand, Tudor had to make natural ice both convenient and affordable. He had a lot of trouble achieving both these objectives simultaneously.

Complete unfamiliarity with the article rather than prejudice against ice formed the most significant barrier to ice adoption in tropical climes. "Ice," wrote Tudor in 1826, "when first introduced to a population unaccustomed to it, is a nine days wonder and after that is over, they will very few of them sit about to use an article to which they are totally unaccustomed and to most of them is a mere sight of something which they have never seen before."[31] During Tudor's first venture in Martinique, one customer took his ice home and left it out in the sun in front of his door on a plate. Later the customer returned to Tudor and complained that it melted.[32] Such missteps did not faze Tudor because he believed that he could cultivate a taste for ice among his customers "in *a course of years*." Unfortunately, as Tudor recognized, "*a course of years* [would] inevitably be fatal" to his business.[33] That explains why he went to such great lengths to accelerate the rate at which people adopted his product.

While less scrupulous businessmen might have let people buy ice only to see it melt away, Tudor wanted to create permanent customers. To keep customers coming back for more, Tudor gave them storage containers of his own invention, and these receptacles formed a crucial piece of Tudor's rudimentary cold chain. The first version of these receptacles, a fourteen-gallon jar surrounded by two inches of sawdust and a layer of dry moss, worked well enough for demand to develop. It had two openings, one at the top where the ice would be added and another below where water drained out.[34] Tudor gave these ice canteens mostly to saloon owners from Charleston to Rio de Janeiro so that they would have more ice to distribute. Despite protestations from some of his agents who thought this an extravagant expense, Tudor persisted in this strategy. "They are a very great convenience," he explained to one employee, "separate from their economy, for barkeeps and gentry of that sort lower down. They pay for themselves in one year and last as many years as are taken care of."[35] By helping consumers preserve their ice, Tudor could convince people of its value, since they would enjoy it more if it lasted longer. As people learned how to store ice for personal consumption, demand for the product increased. In good seasons, Boston ice arrived in the southern ports of Charleston and New Orleans. Cities on the eastern seaboard like Philadelphia and Baltimore also came to look to New England for thick, high-quality ice even though it sold at a higher price.[36] Intense demand even drove up the price of sawdust to insulate ice for shipping.[37]

Tudor also tried to increase demand by getting more people to buy ice in the first place. "A man who has drank his drinks cold at the same expense for one week," Tudor wrote his icehouse keeper in Martinique in 1820, "can never be persuaded to drink them warm again."[38] To get more customers to demand ice with their drinks, he gave his product away free for a year in all the ports where he wanted to cultivate a natural ice market. In the hard-drinking culture of early-nineteenth-century America, the patrons of saloons constituted the primary market for his product. In 1821, Tudor told his agent in New Orleans to pick four such places frequented by "low people and negroes." Then he instructed him to make a confidential bargain, giving those establishments their ice for free for an entire season on condition that they charged the same price for cold and tepid drinks alike. Once he had customers hooked, Tudor increased the price of ice to whatever the market allowed.[39] He reasoned that if customers saw ice as a necessity rather than a luxury, they would pay any price for it. This did not happen during Tudor's lifetime. Instead, demand for Tudor's ice rose and fell with the rest of the economy. The decline in ice purchases during hard times such as during the Panic of 1820 was what led to Tudor's multiple stints in debtor's prison early in his career.

The entrance of new firms into the industry signaled that Tudor's efforts to develop a market for his product had succeeded. His first serious competitors (as opposed to people experimenting with ice shipments or carrying ice as ballast) appeared during the early 1830s, when a number of other Boston firms entered the trade. The most prominent of these competitors was Gage, Hittinger & Company, which changed its name repeatedly until folding in 1889.[40] As competition grew, Tudor's share of the business shrank. In 1849, just as his ice ventures began to create a stable fortune for him, Tudor estimated that he controlled no more than a quarter of the trade in Boston.[41] Once he had demonstrated the feasibility of transporting ice to the tropics, any vessel could pick some up in Boston and try to sell it at points south. Even if these ships could not sell the ice they packed, it still served as ballast on journeys where the hold might have otherwise gone empty. Most of his competitors did not stay in the business long though, because, unlike Tudor, they could not make the natural ice trade profitable.[42]

Still, even these short-lived competitors managed to change the industry. Tudor's rivals slavishly copied his innovations, including the use of Wyeth's ice plow, thereby increasing the efficiency of their operations. They also competed with Tudor to secure waterfront property throughout New England, which included the right to harvest ice on whatever body it adjoined. In 1838, Tudor bought a large farm along the shores of Fresh Pond. He did so not just to ensure access to the shores of the lake to cut ice but also so that he could run the farm to augment his income. In 1840, one of Tudor's competitors built a considerably larger icehouse along a part of the shore that Tudor did not control. Recognizing a legal quandary over ice cutting rights on pond land that they did not technically own, all ten of the ice harvesting concerns that owned land along the shores of the pond appointed an arbiter to divide the pond. He did so on the basis of the size of their land holdings along the shore. This method became the standard for resolving competing ice-cutting claims throughout the country.[43]

The growth of the industry over time also demonstrated Tudor's marketing success. In 1806, Tudor's one cargo leaving Boston amounted to 130 tons of ice. By 1816, six cargoes leaving Boston amounted to 1,200 tons; and by 1836, 45 cargoes carried 12,000 tons.[44] "More ice is now secured in one favourable day," reported the *Edinburgh Journal* about the American ice trade in 1849, "than would have supplied the whole trade in 1832."[45] By 1880, the total ice consumption in the United States amounted to between 5 million and 5.25 million tons of ice each year.[46] Frederic Tudor died in 1864. The total amount of ice harvested in the United States rose every year from Tudor's death to the end of the nineteenth century.[47] Despite the emergence of mechanical refrigeration, natural ice consumption continued to grow into the new century as more and more people developed a taste for ice of any kind.

Although impressive, these growth numbers obscure an important transformation in the market for natural ice. By the time Tudor died, the long-distance transport of ice had become the exception rather than the rule. By 1880, ice merchants no longer had to ship their product to tropical climates at all in order to sell it because Americans in colder climates had started buying more ice, even in winter. For example, between 1847 and 1854, while ice exports were still growing, local consumption around Boston grew from 27,000 to 60,000 tons.[48] Invariably, ice that might have gone overseas to tropical climes got diverted to Boston because the shorter cold chain meant less risk and more ice left to sell.

In addition, ice merchants found that the overseas markets that Frederic Tudor had cultivated became unprofitable over the long run. The explanation for that was both cultural and logistical. India and Great Britain, two markets for natural ice that received enormous attention in the mid-nineteenth century, illustrate these reasons well.

India and Great Britain

The ice trade to India did not begin on Tudor's initiative. Another Boston merchant, Samuel Austin, already traded with India but needed something marketable to transport from Boston to India as ballast in order to make his trade profitable. Tudor had wanted to sell ice in India for years, but did not risk it until Austin came along to help pay the costs.[49] Tudor's first ice shipment arrived in Calcutta in 1833. This trade soon outstripped the rest of his operations. In the 27 years before opening up the market to India, Tudor had exported 4,500 tons of ice from Boston to points south.[50] When the trade to India peaked in 1856, Tudor and his competitors around Boston sent 146,000 tons of ice there in a single year. (This proved to be the peak year for American ice exports over the whole history of this industry.)[51]

Around the time that the Indian market opened, Tudor realized that he could no longer undercut prices to keep rivals out of his many markets because of the distance involved. As a result of that competition, the number of ships leaving Boston with ice escalated sharply. The Indian market had much to do with this increase in volume. Tudor had to ship more ice there, since more of it invariably melted along the way than would have happened had the ice gone to closer ports.

Tudor had no trouble finding people in India who wanted ice to seek relief from the heat. "Like most other conveniences which habit renders familiar, the ice has almost ceased to be regarded as a luxury," wrote the British artist Colesworthy Grant from Calcutta in a letter to his mother back in London, "and though little children still continue to seek and to suck it as though it were a sweetmeat, they no longer consider it as the novelty which, when first holding it in their nearly paralyzed little fingers, they declared, in amazement, had *burnt* them!"[52] While this observation might suggest India's suitability as a market for ice from New England, the business end of the enterprise proved much more complicated. Indian natives did not earn enough money to buy ice regularly. British colonials did—but not every British colonial: only the kind who had enough money to hire servants could afford top buy ice. One of twelve classes of household servants in India, the *abdars*, had had the job to preserve perishable foods and provide their employers with fresh water. That job disappeared once American ice arrived in India.[53] Yet the purchases of this small class of people never justified the expense of sending ice around the world from New England.

Tudor's initial shipment to Calcutta arrived to a rapturous reception. "The first transport of Ice, from the shores of the United States to the banks of the Ganges," wrote the *Calcutta Courier* in 1837, "is an event of no mean importance; and the names of those who planned and have successfully carried through their adventure at their own cost, deserve to be handed down to posterity with the name of other benefactors of mankind."[54] The people of Calcutta eventually took up a collection to erect large icehouse at their own expense to better store their cold comfort. Had New Orleans or Charleston ever responded this way, Tudor never would have had to worry about his finances. Unfortunately, recovering costs from Indian shipments became very difficult. By 1870, six years after Tudor's death, his ice company made a profit only in Calcutta, while the markets in Bombay and Madras produced large losses despite many government-issued privileges.[55] Citizens in all three of the Indian ports also built large icehouses and leased them to his company at low rents.[56] Because of these difficulties, Tudor had to continue to dabble in other lines of business to stay economically afloat. For this reason the India trade, to a much greater extent than other routes, included foodstuffs along with the ice.

In 1850 Gage, Hittinger & Company sent three shipments of ice to India in order to break into a market that Tudor had mostly to himself. This company sold its product at half of Tudor's price. In response, Tudor cut his prices half again below Gage, Hittinger's price, something he obviously preferred not to do.[57] This decision to start a competition in India probably came about because Gage, Hittinger had failed to cultivate the British market, which it pioneered by sending ice to London in 1842. In 1845, one London newspaper noted that "no banquet of any magnitude is considered complete" without ice.[58] In an approach like the one Tudor used in the tropics, Gage, Hittinger tried to stimulate demand for a product that few people in England knew or understood. Most notably, the company sent several skilled bartenders to the island to teach the English how to make American-style mixed drinks. By 1863, England consumed between 50,000 and 60,000 tons of ice per year.[59] Despite this apparent success, the ice market in England ultimately failed due to a combination of logistics and culture.

Charles Lander and Henry T. Ropes of Boston proved somewhat more successful when they founded the Wenham Lake Ice Company, named after their source in Massachusetts, shortly after Gage, Hittinger opened up the English ice market. Marketing became the most important factor in the early success of Wenham Lake ice. A block of ice placed in the company's office window in London's Strand district attracted enormous attention from passers-by because it appeared indestructible, when, in fact, the company merely replaced it frequently. The company scored another marketing coup when it dispatched a block of ice to Queen Victoria. With her endorsement and subscription for a regular supply, the market grew quickly—but only among the English aristocracy.[60] Ice, as Sarah Maury explained in 1848, "was a luxury only indulged in on occasions of company and display, but among the middle classes it was unknown."[61] Because of a small market and high transportation costs, the price of ice did not come down fast enough in England to create American-style demand. In fact, fear of the health consequences of ice might have meant that most Britons never consumed any.

Since the English poor and middle classes could not afford such a luxury, it did not change the way Englishmen ate and drank. Instead, by far most of the ice entering England went towards industrial uses, mostly preserving fish. Around 1850, the proprietors of the Wenham Lake Ice Company switched their source from Massachusetts to Norway, where they used the same harvesting methods, in order to save money. Since Norwegian lakes were also clear and blue like Wenham Lake, the firm continued to call their product "Wenham Lake Ice," and English customers never knew the difference.[62] The American ice trade to England ceased around 1880, completely replaced by imports from Norway at a lower price.[63]

The natural ice trade from Norway to Great Britain actually predated the trade from the United States, beginning in 1822. The customs officers at the time had no idea of how to classify ice for payment of duties, so it often melted before reaching its final destination.[64] The size of the Norwegian ice trade increased remarkably after a few representatives of the Norwegian industry visited America during the 1840s to learn about American innovations (such as ice tools and plows) in order to copy them. The British-Norwegian ice trade surpassed the trade with America between 1850 and 1865. The great leap in efficiency that made the price of Norwegian ice go down below American levels came about because of the construction of channels that allowed lake ice to travel rapidly down hillsides directly to portside, where the ships to take it away awaited, thereby shortening the cold chain on land as well as on water. As a result of all these improvements, Norway shipped ice to New York City profitably in 1880, the year of a great ice famine in that city.[65] The Norwegian trade peaked at the end of the nineteenth century, with approximately 2.2 million tons making the trip across the North Sea between 1893 and 1898; after the turn of the century, ice was made in England by machine.[66] While the cold chain practically disappeared in Great Britain, shorter cold chains in America more than made up the demand lost from overseas.

Local Cold Chains along the Hudson and Elsewhere

By 1880, the American natural ice industry had come of age. "No man sows, yet many men reap a harvest from the Hudson," wrote the naturalist John Burroughs about the New York ice trade that same year, keeping up Thoreau's tradition of comparing ice harvesting to agriculture. "Ice or no ice sometimes means bread or no bread to scores of families, and it means added or diminished comfort to many more."[67] Ice harvesting around New York City had begun in 1816.[68] The scale of activity had increased greatly by the time that Burroughs wrote about the trade:

> The cutting and gathering of ice enlivens these broad, white desolate fields amazingly. My house happens to stand where I look down upon the busy scene. . . . Sometimes nearly 200 men and boys, with numerous horses, are at work at once, marking plows, planing, scraping, hauling, chiseling; some floating down the pond on great square islands towed by the horse, or their fellow workmen; others distributed along the canal, bending to their ice-hooks; others upon their bridges, separating the blocks with chisel-bars; others feeding the elevators; while knots and straggling lines of idlers here and there look on in cold discontent, unable to get a job.[69]

In 1882, approximately 135 large icehouses stored the product harvested between New York City and Albany. Most of them could hold between 20,000 and 85,000 tons of ice. Many more smaller ones that held between 3,000 and 5,000 tons also lined the banks of the Hudson.[70]

The mechanization of at least part of the harvesting process made this large-scale process much more efficient than earlier in the century. Foundry owner Elihu Gifford invented the first conveyors designed to move natural ice directly from the water into the icehouse around 1840.[71] By late in the century, these conveyors had become gigantic ice elevators that could move many tons of product into increasingly immense storage spaces. These conveyor belts (called "endless chains") stretched from the edge of the lake to the top of huge ice houses. "When all works well," wrote Burroughs, "there is an unbroken procession of the great crystal blocks slowly ascending this incline. They go up in couples, arm in arm, as it were, like friends up a stairway, glowing and changing in the sun. . . . When they reach the platform where they leave the elevator, they seem to step off like things of life and volition."[72] A 40-horsepower steam engine employed to power this machinery could do the work of 150 men and 75 horses faster and more efficiently.[73] Imagine the number of men and horses needed to do the same work back on the ice. Then factor in the increase in the scale of operations even just from Thoreau's time along Walden Pond, and it becomes possible to fully appreciate the improved efficiency of these later operations.

With ice consumed closer to its point of origin, ice merchants no longer worried so much about how well their storage slowed melting. Huge blocks of ice stacked in gigantic wooden shacks kept themselves cold (or at least the parts that were not exposed to the air did). Typical icehouses along the Hudson and in other areas with large ice harvests measured between 100 and 150 feet long and were as much as 40 feet wide in this era. The companies separated these huge structures into "apartments" divided by wooden walls to make loading the inner parts of the houses easier.[74] By 1880, the seventy-four icehouses along the banks of the river could hold two million tons of frozen water. The companies whitewashed these buildings on the outside to reflect the heat of the sun on clear days. Inside the houses, sawdust or straw separated each block and served as insulation.[75] Ironically, the insulation made icehouses fire traps, prone to burn in lightning storms and therefore almost impossible to insure. Here, as before in New England, the ice cut in winter sat until summer. Once summer arrived, boats came up the river to transport the cargo to the suppliers' markets of choice. While Tudor's ice had traveled into the torrid zones of the Americas and the world, Hudson River ice usually traveled no further than the metropolis at the mouth of the river to quench the thirst of the urban masses there. New York City consumed almost 1.9 million tons of ice in 1882, most of it delivered down the Hudson by barge.[76] Sometimes Hudson River ice went as far west as Cincinnati.[77]

Without the development of local markets and the shortening of the cold chain, this increase in the scale of harvesting operations never would have made a profit. Greater production required greater consumption. A better survival rate for the harvested ice greatly increased the overall supply. Shorter distances to cover meant more ice to sell because less of it melted on its way to market. Increased supply generally meant lower prices, and that meant even the poor could afford to buy it. These short, efficient cold chains also allowed ice dealers to specialize in selling ice (unlike Tudor, who sold food too), since they did not have to charter a ship to carry tons in cargo or to fill it with different goods on the voyage back to its home port in order to make up for that cost.

Unfortunately for the natural ice industry, the weather in the Hudson Valley did not always cooperate with their interests. What the industry called "open winters" happened when the temperature stayed too warm to form harvestable ice. Such situations made the ice business around the country very volatile. Ice "famines" struck the Hudson in one of three winters during this period. In 1880, for example, the weather stayed so warm that the New York harvesters cut only 600,000 tons of soft, spongy ice that winter.[78] This led to the rapid growth of ice harvesting in Maine.

The appeal of Maine ice came from the character of the water there. The Kennebec River, the center of the state's industry, originates from underground springs in northern Maine. Before the sawmills came, the quality of the water was good, and the turbulence of the fast-running river helped keep sediment from accumulating so that Kennebec ice was clear and blue.[79] People loved dropping this ice in drinks because it left a minimal amount of sediment behind. By 1890, the ice crop in Maine had increased to over 3 million tons in response to yet another open winter on the Hudson. This led to a veritable ice rush. As the price of ice skyrocketed, the volume of ice harvested in the state increased by 1.5 million tons over the previous record level.[80] Maine continued to produce natural ice at a similar rate for the next twenty years. When summer came, barges headed for New York, Philadelphia, or Baltimore could go a hundred miles upriver to pick up all the ice that had been cut the previous winter and drop it off directly at those cities' proverbial doorsteps.[81]

At the same time that Maine began to supply East Coast cities, a significant ice industry gradually arose in the Midwest and in California. These markets were mostly local. Only Maine ice was sold outside its immediate region. Most markets simply took the ice closest to them because they could then sell it at the cheapest possible price. The cold chain began to contract before mechanical refrigeration made it possible for it to expand again.

The year 1880 marked the apex of the American natural ice industry. Volume increased after that year, helped along by new innovations that improved the efficiency of harvesting, transporting, and storing the ice. However, the importance of this product to Americans diminished as another form of ice gradually became viable in its own right. Having two kinds of ice available also led to new terminology. Ice cut from lakes or rivers became known as "natural ice." Ice made by machine became "artificial" or "mechanical" ice.

With a superior alternative available, natural ice became increasingly less desirable to most consumers. When the onetime novelty of machine-made ice became commonplace, convenience and aesthetics gradually became more important to consumers of cold than simply the fact that they could control the cold at all. Natural ice persisted longer as a component of the cold chain rather than as the product shipped through the cold chain. Aesthetics did not matter for these kinds of uses; simply having a regular supply of cold did. Over time, as ice machines became more reliable and the price of machine-made ice dropped, natural ice lost this market too.

The large, expensive machines for ice making did not compete directly with ice cut from lakes and rivers until the 1880s, when the natural ice industry faced particularly bad weather and high demand that it could not meet. But the natural ice industry did not disappear overnight. Instead, a particularly brutal battle between the natural and mechanical ice technologies to determine dominance ensued over a span of decades. Because it took many years to perfect mechanical refrigeration on a large scale, even more to make it useful for the transportation and storage of food, and still more to shrink this technology for home use, natural ice, a quintessential nineteenth-century product, lasted well into the twentieth century, filling in places along the cold chain until mechanical refrigeration could fit in the gap.

CHAPTER 2

The Long Wait for Mechanical Refrigeration

On the afternoon of 10 July 1893, Captain James Fitzpatrick of the Chicago Fire Department received yet another call to put out a fire in the ice-making and cold storage facility at the World's Columbian Exposition. He had gone to fight fires there twice before. A design defect in the plant's smokestack had caused each of those blazes.[1] To obscure the stack from the view of fairgoers, the architect had designed a 225-foot wooden encasement tower that stood 5 feet taller than the stack itself. According to the architect's plan, an iron "thimble" went atop the smokestack to prevent particles and debris from igniting the wooden tower outside. For whatever reason, the company that constructed the fair's ice-making machinery and the building that contained it never installed the thimble. Newspaper accounts of the July blaze report that it started when flames from soft, greasy-burning coal used to fire the boiler below ignited soot in the upper reaches of the smokestack.

When Captain Fitzpatrick arrived at the scene, he assumed that the blaze resembled the ones he had faced earlier. As he did on these other occasions, Fitzpatrick ordered his men to use their ladders to climb the outside of the tower and fight the fire where it had started. "We'll put this blaze out in a minute," he said at the time.[2] Unbeknownst to Fitzpatrick, the fire had already spread from the top of the smokestack to the building below. Fitzpatrick and his men alighted onto a balcony approximately fifteen feet below the flames. Shortly thereafter, the building below them exploded, sealing off their escape path. According to one eyewitness,

> Immediate confusion followed among those on the balcony. Two men at once jumped into open space, two were lost on the west side of the tower, and the balance of the frightened human beings rushed to the north side of the tower and huddled together, all-fearful of the oncoming flames and apparently realizing that they were now between two deaths—that of being burned or crushed [by the falling tower]. Desperation probably caused them to seek the latter, and out into the air in rapid succession shot a half-dozen human beings, whirling and circling over

> and over. When these poor men struck the flat roof eight feet below they bounded back into the air and fell back again to struggle with death. At last but two men were left surrounded within five feet on all sides with fierce flames. It either was to jump quick or death from fire. One grabbed a rope, started slowly down, passed through two sheets, and then the rope burned in two and the hanging men fell to the roof, bounded into the air a confused mass of arms and legs, and again fell back. As the next man took hold of the rope the four walls separated like melting crust and he was whirled into the burning tower, while the east wall, covered with fire, fell on the man who jumped before him.

Thousands of fair visitors watched the tragedy as it happened. As each firefighter fell, "a simultaneous murmur of horror escaped from throats for fully a half mile in every direction."[3] The blaze killed seventeen people in total, including Captain Fitzpatrick and eight other firemen. It left nineteen people, including five firemen, seriously injured.

The ensuing investigation of the fire focused on the faulty smokestack. The legal system never determined ultimate responsibility for this fatal flaw.[4] Regardless of how the fire started, the firemen could have extinguished the blaze in the encasement tower without loss of life but for the explosion, just as they had twice before. What then made the building explode? The *Chicago Inter Ocean* blamed "ammonia in the refrigeration system," which once it caught fire led to a "deafening explosion of the ammonia pipes."[5] While nobody knows the exact cause of this fire, informed observers and many people inside the refrigeration industry believed that an ammonia explosion caused the Columbian Exposition disaster. This belief slowed the widespread adoption of ammonia compression refrigeration technology in the United States.

Despite the perceived risks that this tragedy represented, many American companies nonetheless tried an unproven technology that they did not entirely understand because it had many advantages over natural ice both for producing and managing the cold. Natural ice offered temporary control over cold, but natural ice always melted eventually, and it could not bring the temperature of anything below the freezing point. Natural ice also required an extensive infrastructure to harvest and transport, which could make it expensive. Once perfected, mechanical refrigeration produced reliable cold at the point where businesses needed it. As the technology improved, it became possible for operators to bring a space to a precise temperature, an important requirement for successful cold storage. Mechanical refrigeration required a large up-front cost, but little additional money after that except for the cost of the power to keep the machine running. For all these reasons, the creation of the modern cold chain depended upon the development of reliable mechanical refrigeration.

Because of the problems associated with natural ice, inventors had experimented with the components of mechanical refrigeration as far back as the seventeenth century. Refrigerating engineers had many choices besides ammonia to serve as possible refrigerants in the late nineteenth century. The search for the most efficient, reliable, and safe refrigerant formed one part of the larger effort to create efficient, reliable, and safe mechanical refrigeration. That search took the better part of the nineteenth century to complete. The first refrigeration machines weighed tons. Because they were powered by steam, only warehouse-sized systems could refrigerate enough of anything to make building these machines profitable. As with the perceived safety risks, such difficulties greatly slowed the widespread adoption of mechanical refrigeration. Experts in the field saw the potential of this technology, but it took decades to perfect it.

As refrigerating engineers in the United States and Europe tackled the problems associated with mechanical refrigeration, the equipment not only shrank; it became reliable enough for businesses to run it profitably on a regular basis. Effective mechanical refrigeration changed the way people lived and, perhaps to greater effect in the long term, how they ate. In the short run, mechanical refrigeration lowered the price of ice in places where it seldom (if ever) developed naturally during the wintertime. Lower prices made ice more popular in hot and dry regions and gradually led more people everywhere to buy it regularly.

Ice machines had serious drawbacks in their early years, but these problems did not halt the growth of the mechanical refrigeration market. The men whose business depended upon refrigerating machinery found many models unreliable, difficult to operate, and almost prohibitively expensive because of their huge size. While many owners ignored the safety issues associated with this equipment that the Chicago fire and other less-publicized tragedies brought to public attention, those issues nonetheless lingered in the public mind well past the moment when refrigeration equipment became safe. As refrigerating machines improved, they became popular for many industrial purposes, from making steel to preserving dead bodies. As time passed, they also became increasingly popular around the world, thereby making the first efficient global cold chains possible.

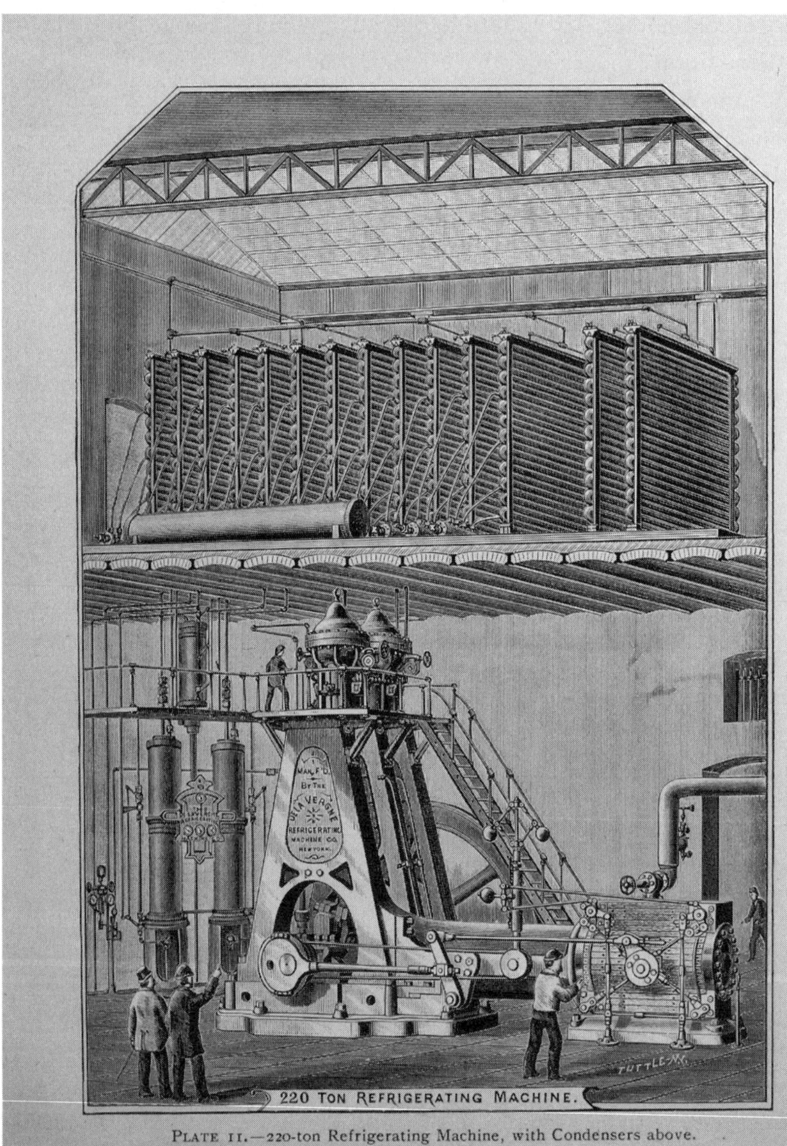

This "220-Ton Refrigerating Machine" from the De La Vergne Refrigerating Machine Company's 1890 catalog suggests the huge size and weight of mechanical refrigeration equipment in the early days of this technology. Courtesy of the Hagley Museum and Library.

The Principles of Mechanical Refrigeration

In 1883, Mark Twain published the memoir *Life on the Mississippi*, a book that simultaneously considered his past as a steamboat pilot before the Civil War and the changes along the river between the end of that war and the time of the book's publication. Perhaps because of his former life on the water, Twain paid careful attention to ice both on and off the river. "In Vicksburg and Natchez, in my time, ice was jewelry," he wrote; "none but the rich could wear it. But anybody and everybody can have it now." The development of artificial ice made the difference on this front. During a return visit to New Orleans, Twain visited an artificial ice factory there "to see what the polar regions might look like when lugged into the edge of the tropics." The place did not impress him. To Twain, "it was merely a spacious house, with some innocent steam machinery at one end of it and some big porcelain pipes running here and there." Mechanical refrigeration may have appeared ordinary, but the fact that Twain did not understand how it worked suggests the complexity of this technology.[6] The production process for creating clear ice remained opaque to him.

Many people shared Twain's confusion about how mechanical refrigeration worked. Only as the principles of mechanical refrigeration became widely understood did the industry come into its own. The fair in Chicago offered the industry a chance to expand public understanding. In December 1892, after the announcement of the contract to build the ice-making and cold storage pavilion at the Columbian Exposition but before the ultimately doomed structure was built, the trade journal *Ice and Refrigeration* lamented that the fair's organizers did not take the technology of the industries they covered seriously enough. "Electricity, transportation, mines, fisheries, other industries, have entire and enormous buildings provided by the Exposition company," it complained, "but refrigeration, that great science which is revolutionizing the production, conservation and distribution of food . . . is buried."[7] Under pressure from the refrigeration industry, the organizers of the Columbian Exposition agreed to let the Hercules Iron Works include its own display about refrigeration inside its building. Unfortunately, that exhibit burned up in the fire.[8]

People who should have known more about refrigeration were just as ignorant about it as the general public was. In its report on a meeting of the National Association of Stationary Engineers, an 1894 *Milwaukee Journal* article described Otto Jahn as "an ice expert." Jahn, however, explained his limitations with regard to sharing that expertise. "If I were to explain the principle and

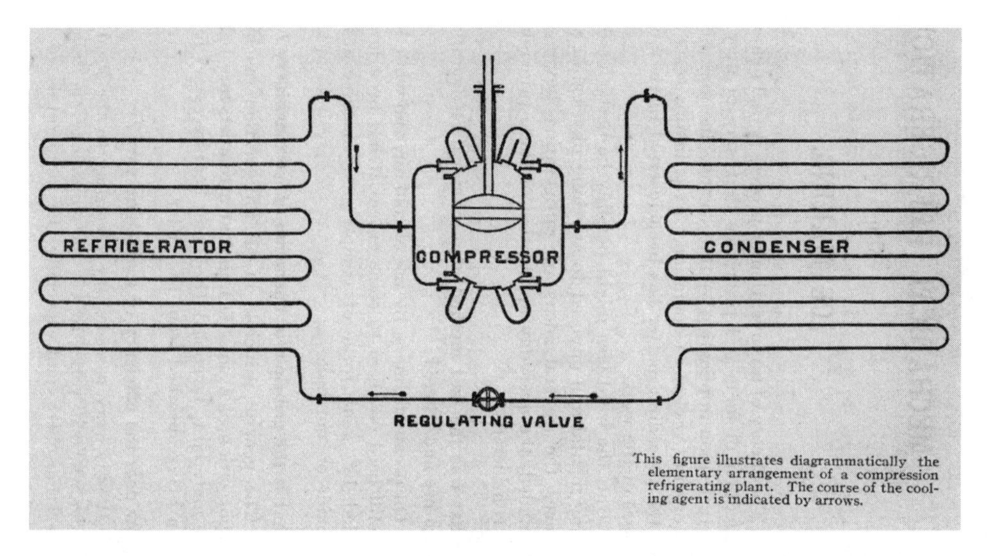

The compression refrigeration cycle (compression, evaporation, and expansion) applied to machinery that used any kind of refrigerant during the early days of the industry just as it still applies to compression refrigeration machinery in use today. From Gideon Harris and Associates, *Audel's Answers on Refrigeration and Ice Making* (New York: Theo. Audel & Co., 1914).

construction of all the various refrigeration machines on the market, with full directions as to how to operate them in the most practicable and economical way," he told his colleagues, "it would require a few hundred nights so far as time is concerned."[9] The difficulties that such experts faced in communicating how refrigeration worked explains as much about their relative ignorance as it does about that of their audience.

Fairgoers who visited the ice and cold storage building would have learned basic principles of refrigeration that still apply today. While the machinery has become much more complex in the more than one hundred years since that time, refrigeration principles have not changed. Technically, cold is the absence of heat, not a force in its own right. Thus, the process of refrigeration or cold storage requires use of a substance that will draw heat away from the area where one wishes to lower the temperature. Chemicals that can perform this function are called refrigerants. A refrigerant "acts like a sponge," explained one refrigerating equipment company to its potential customers. "It sops up heat in one place and parts with it in another."[10] The best refrigerants have the highest latent heat, which is the amount of heat needed for to vaporize a substance after it reaches its boiling point.[11] Because the relationship between the temperature, volume, and pressure of gases is fixed, volume and pressure can be altered to manipulate the refrigerant's temperature. To do this, most refrigerating machinery puts the refrigerant through three stages: compression, evaporation and expansion. These three processes must occur in any compression refrigeration system, no matter what time period one considers.

In the first and most important step of artificial cooling, the refrigerant in its gaseous state gets compressed into a smaller space. In the early days of mechanical refrigeration, steam engines usually powered compressors.[12] Later on, power came from electricity. Either way, one stroke of a piston draws gas in from the evaporating coil and the opposite stroke compresses it.[13] During compression, heat is generated in proportion to the amount of pressure put upon the gas. The hot compressed gas is then forced through a condenser of coiled pipes submerged in cold water until enough heat is drawn off to liquefy the refrigerant. At this point, the now-liquefied refrigerant is drawn through an expansion valve. The expansion valve lowers the pressure to a point at which the refrigerant returns to a gaseous state. This liquid-to-gas transformation requires heat. Heat drawn from the surroundings is extracted from the gaseous refrigerant in the condenser and absorbed by the liquid refrigerant in the evaporator. The resulting removal of heat in the surrounding area lowers the temperature outside the machine, thereby making the desired space cold. Owing to the expense of efficient refrigerants, all refrigerating machines operate as closed systems, recycling the gas and liquid so that they can undergo compression and expansion over and over again.

Those who experimented with refrigeration technology during the early nineteenth century were usually wanting to produce ice. Ice machines and refrigerators used essentially the same technology, with ice machines chilling water and refrigerators chilling space. Therefore, the development paths for these inventions overlapped. Working in London, the American inventor Jacob Perkins created the first working refrigeration system in 1834, utilizing ether as the refrigerant. The system included all the principal parts of the modern compression machine: the compressor, the condenser, the expansion valve, and the evaporator. Perkins's machine could actually make small quantities of ice, but with no demand for ice in Great Britain at the time it attracted no attention.[14] Professor Alexander Twining of Yale University, in a series of experiments begun in 1847, improved on this ether compression technology and made it commercially available.[15] He patented his system in England in 1850 and America in 1853. A machine Twining installed in Cleveland produced 1,600 pounds of ice per day. Unfortunately, since Cleveland (and, in fact, much of America) could easily get cheap natural ice, Twining's machine could not compete with the local product. The same system was used more often in Europe, particularly in France between 1857 and 1864, but Twining made no money from these efforts, as he did not authorize them.[16]

In 1851, a Florida doctor named John Gorrie patented a new way to make ice with machinery that used air as the refrigerant.[17] He discovered this process accidentally in the course of inventing an ancestor of modern air conditioning for his yellow fever patients when his cool air machine "iced up."[18] Few people believed that he had produced ice this way. Despite the doubters, Gorrie developed his ice machine in New Orleans and Cincinnati and licensed it overseas, but his efforts also failed commercially.[19] In 1869, the German engineer Franz Windhausen patented an improved cold air machine along the same lines as Gorrie's. The English firm of J. & E. Hall made further improvements in 1880. This technology proved particularly useful for early experiments in refrigerated shipping, but eventually carbon dioxide (CO_2) systems surpassed it in popularity.[20]

Like cold air machines, machines that used ether as a refrigerant also proved highly inefficient compared with the machines using other refrigerants developed by the end of the nineteenth century. A compressor big enough to create cold from ether needed to be seventeen times larger than a comparable ammonia compressor, the most efficient refrigerant available by the end of that century.[21] Ether compression alone, therefore, would never have met the considerable demand for ice that developed during this era. Ether is also highly flammable. Finding a safer refrigerant proved absolutely essential to the development of the industry. The fire at the Colombian Exposition in Chicago tragically illustrated that ammonia had risks of its own, since gaseous ammonia mixed with air can deflagrate between the levels of 16 and 25 percent. Deflagration is an intense shock wave that resembles an explosion. Mixed with oil, ammonia can conflagrate at a broader range of percentages.[22]

Ice equipment manufacturers understood this possibility before the tragedy in Chicago.[23] Ammonia compression refrigeration systems became popular despite these safety concerns because the industry cared more about efficiency than safety.[24] "What bothered us most was the production of any kind of cold in regular quantities," remembered Thomas Shipley of the York Company, an important ammonia equipment manufacturing firm. "What we were after was cold weather and as much of it as we could get."[25] The consumers of refrigerating equipment—the owners of artificial ice companies and cold storage warehouses scattered across the United States—almost always shared these priorities.[26] Firms in other countries (and a few in America) mostly made less-efficient choices for the refrigerant in their refrigeration systems because they cared more about safety.

Four Refrigerants

For most of the nineteenth century, inventors and firms had trouble deciding on the best refrigerant for mechanical refrigeration. In 1931, a Du Pont engineer named W. D. Humphrey estimated that some fifty different substances or mixtures had been used at one time or another as refrigerants.[27] Some of these substances proved better than others at their appointed task. The four refrigerants that worked best—ammonia, aqua ammonia (ammonia mixed with water), sulfur dioxide, and carbon dioxide—formed the basis of almost all practical refrigerating machines during the nineteenth century.[28] The refrigerant that was most effective depended on the purpose for which it was used, but discovering what refrigerants worked best for particular objectives took many decades of trial-and-error experimentation. Determining the best refrigerant for use in different kinds of transport and storage proved crucial for making the modern cold chain possible.

In 1859, the French inventor Ferdinand Carré created the first refrigeration machine that used aqua ammonia as the refrigerant. Refrigeration systems based on aqua ammonia worked somewhat differently than later closed ammonia compression systems. Refrigerating engineers later called the aqua ammonia system an ammonia absorption refrigeration system. Carré used a boiler to heat aqua ammonia to produce ammonia gas and pressure it to the point of liquefaction as the gas traveled through pipes. Cold water surrounding the pipes drew off heat from this operation. The system then reexpanded the liquid into a gas in coils surrounding the substance to be cooled. Finally, the gas ran back into the ammonia-water mixture, where it got absorbed back into the system because of ammonia's high affinity for water.[29] Because the ammonia and water molecules remained locked together during much of this process, this system posed very little fire risk.

Ammonia absorption machines debuted in the southern United States during the Civil War. They were exported from France, by shippers who ran the Northern blockade of the South. While this kind of machine had many drawbacks, the South had nowhere else to turn with shipments of natural ice cut off for the duration of the conflict.[30] Because there were no viable alternatives, ammonia absorption systems remained the most popular in the world until 1875. They served as the impetus for the beef export industry in Australia and South America before the perfection of ammonia compression systems in the late 1870s.[31] This technology remained in America even after alternatives emerged, maintaining a 12 percent market share in the United States as late as 1919.[32]

In 1874, Raoul Pictet, professor of physics at the University of Geneva in Switzerland, produced the first of a series of compression refrigeration machines that used sulfur dioxide as the refrigerant. Sulfur dioxide is a nonexplosive liquid at room temperature. Furthermore, its oily nature eliminated the need for lubrication in the machinery. (When oil found its way into ammonia pipes, and it inevitably did, the efficiency of the ammonia compression system decreased.) The absence of an oil lubricating apparatus simplified the mechanical details of sulfur dioxide refrigerating equipment, making the machine easier to run and less likely to break down. Sulfur dioxide, however, has a lower latent heat than some other refrigerants used at this time, and it carries the added disadvantage of smelling very bad, with a kind of a musty odor at low concentrations and a very acidic smell at high concentrations.[33] It is poisonous, but in this regard its bad smell was actually an advantage, since potential victims would have smelled leaks and evacuated the premises where the machine was located long before getting poisoned. Despite these drawbacks, Pictet installed a number of machines in France and a few in the United States.[34] France continued to use sulfur dioxide machines well into the twentieth century.

During the 1880s, Windhausen switched from air cycle to carbon dioxide refrigeration systems because carbon dioxide proved generally more efficient and more reliable.[35] These machines, which subjected carbon dioxide to the same kinds of processes used with ammonia in ammonia compression machines, had some advantages over the ammonia refrigeration systems. Unlike ammonia, carbon dioxide does not corrode copper. Carbon dioxide machines became very popular in ships because their refrigerant was nontoxic and because the copper pipe used with them did not rust in the sea air.[36] Carbon dioxide also cost less than ammonia did. At the turn of the twentieth century, carbon dioxide cost approximately five to seven cents per pound, since industrial carbon dioxide was a by-product of making beer.[37] Therefore, operators using the carbon dioxide compression system paid substantially less for their refrigerant even though ammonia has the higher latent heat of these two substances. Besides being cheaper, carbon dioxide was nonflammable and nonirritating. Granted, the high pressures required for carbon dioxide created the possibility for more leaks, but the ability to use copper pipe helped this problem. Because of these higher pressures, carbon dioxide machines had much more safety equipment than ammonia compression machines did.[38] Carbon dioxide even extinguishes fires.

Despite its many advantages, the technology for carbon dioxide machines took time to develop. In 1890, one ammonia compression equipment manufacturer estimated that its machinery was eight times as efficient as any machinery that used carbon dioxide.[39] As a result of technological improvements made by J. and E. Hall, carbon dioxide machines became competitive with ammonia starting around that time but gained no more than a toehold in the mechanical refrigeration market until the turn of the twentieth century.

Ammonia (NH_3) is a common household chemical today when mixed with water, which stabilizes the compound. Pure ammonia, however, keeps well only under high pressure and, as explained above, can explode. Despite these drawbacks, many inventors experimented with anhydrous ammonia (meaning not mixed with water) as a refrigerant because its high latent heat suggested that it had great potential for "producing" cold. Both Alexander Twining and the Australian James Harrison took out patents on ammonia compression refrigeration machines during the 1850s, even though they spent most of their time developing ether machines instead. By 1899, 74 percent of American ice manufacturers used ammonia compression machines because, as the journal *Manufacturer and Builder* explained in 1880, "Ammonia is cheap and readily obtainable, and . . . the ammonia machines . . . are not wasteful of power."[40] By 1909, the percentage of American machines that used ammonia as the refrigerant grew to 86 percent.[41] Not until the advent of household refrigerators during the 1920s would firms in the United States experiment with other kinds of refrigeration systems to a significant degree. Most of the American mechanical refrigeration industry got locked into ammonia compression technology. This situation had both costs and benefits.

Professor Carl Linde of Munich Polytechnic College in Germany deserves most of the credit for making anhydrous ammonia a viable refrigerant. Linde's initial research involved the theoretical aspects of thermodynamics, but when he began to develop a relationship with German and Austrian brewers, he started to work out the technical flaws that had prevented ammonia compression refrigeration from living up to its full potential. He installed his first commercial refrigerating equipment in 1877, and he continued to improve his refrigerating machinery for decades afterwards. In 1879 Linde created the Linde Ice Machine Company (a firm that still exists today, although it no longer manufactures refrigerating equipment). As a result of its roots in Linde's scientific laboratory, the Linde Company never built the ice machines it designed by itself. Instead, it farmed the work out to German and foreign manufacturers who bought the rights to build and sell them.[42] The Boyle Ice Machine Company of Chicago, which sold an ammonia compression machine of its own design, went so far as to claim that "machines which use ammonia as the refrigerant are conceded to be the only really satisfactory machines, as they are entirely free from the danger from explosion and fire."[43] This sales pitch was intended to contrast ammonia with ether, but after the tragedy at the Chicago World's Fair in 1893, ammonia's reputation suffered greatly.

"It is not necessary to call the attention of the ice machine operator to the fact that ammonia will explode," wrote the editors of the trade journal *Ice and Refrigeration* in 1894, the year after the explosion in Chicago. "Explosions of this substance are now of not an infrequent occurrence; and as the number of ice-making and refrigeration plants increases from month to month, it is quite possible that the number of reported explosions may increase also."[44] The frequency of ammonia-related fires around the turn of the twentieth century remains a mystery, but insurance records demonstrate that fires often destroyed refrigeration plants and cold storage warehouses as the number of such facilities increased.[45] Before any ammonia fire could happen, an ammonia leak had to occur. Such leaks happened often at ammonia refrigeration plants around the turn of the twentieth century. In a badly run plant, the smell of ammonia permeated the air all the time.[46] One 1919 estimate suggested that 65 percent of the ammonia in any refrigerating plant dissipated through avoidable leaks.[47] A 1920 fire insurance underwriting guide concluded that explosions were frequent in ammonia refrigeration plants because of escaping gas.[48] Yet by 1924, the editors of *Ice and Refrigeration* had ceased to accept the notion that ammonia refrigeration plants could explode, although they did recognize that "whenever an accident occurs in an ice plant, the newspapers immediately attribute it to an ammonia explosion."[49] This explains why the public lost faith in this technology.

The operators of refrigeration equipment, on the other hand, continued to buy ammonia compression machines because traditional ammonia compression systems operated most efficiently and therefore produced the most profits. As long as ammonia remained cheap and the plant did not explode, they accepted the waste of ammonia as an acceptable cost of doing business even if the fumes occasionally annoyed the neighbors. American firms selling carbon dioxide compression equipment made direct comparisons with ammonia in their advertising, paying particular attention to the question of safety.[50] This tactic worked only for a limited group of buyers—in populated places, where safety mattered more. Restaurants, ships and hotels used carbon dioxide machines, but they were used almost nowhere else.[51] In the United States, carbon dioxide refrigeration technology never escaped these niche markets. Most ice companies cared more about production than safety.

In Great Britain, on the other hand, the industry developed primarily for preserving imported food over long distances rather than for making ice. For this reason, mechanical refrigeration equipment buyers there cared much more about safety, both for the food (ammonia would spoil food if it ever leaked, let alone if it blew up) as well as for the people who serviced the cold chain. After licensing the Windhausen patent for carbon dioxide machines, J. and E. Hall conducted pioneering experiments in carbon dioxide refrigeration during the late 1880s. The company's improvements made this technology economically viable and made J. and E. Hall the largest manufacturer of refrigeration equipment in Great Britain.[52] During one of its early installations, a pipe in the system burst because of a mechanical failure. Everard Hesketh, the head of the company later recalled: "The pipe was literally blown to ribbons and had it not been for the sides of the water tank surrounding the compressor . . . my head, which was only a yard away, would have been the target for some of the pieces." Because of this accident, the firm pioneered the use of the automatic safety valve that shut down the system if pressure built up to possibly explosive levels.[53] This innovation greatly spurred the spread of carbon dioxide compression technology outside Great Britain, though not in the United States.

Refrigeration Equipment Makers and Their Customers

The considerable amount of capital required to manufacture and purchase refrigeration machines made mechanical refrigeration expensive compared with natural ice. In most cases (and eventually in all cases as the industry matured), the companies that used ice machines did not build ice machines themselves. Nor did the pioneering engineers who built the first mechanical refrigeration systems in the United States, for they did not have machine shops or foundries at their disposal. As a result, they had to interest a machinery manufacturer of some kind to enter a product line that did not yet exist. That took time.[54]

These manufacturers became the ice machine industry, not the industry that sold ice. While such a division between producers and sellers made the refrigeration industry more complicated, it nonetheless constituted a necessary step for creating a cold chain, as no single company ever grew large enough to take on the risk that vertical integration would have required. As a result, competition grew fierce as the industry expanded, especially after the market became international.

With few funds at an entrepreneur's disposal, the first mechanical ice plants in the United States served as small-scale prototypes. They first appeared in the South during and immediately after the Civil War, since natural ice seldom ever got beyond southern port cities before that time. San Antonio had three artificial ice plants in 1867 when only five others existed in the rest of the United States.[55] American inventors worked on developing ammonia compression machines during the 1860s and 1870s, before Linde had his greatest successes. While these early ammonia compression machines sometimes made profits for their owners, they were not as efficient as those that followed and could not produce large quantities of ice. As the technology became more efficient, the popularity of mechanical refrigeration grew. However, in 1889 there were still only 189 such machines in America.[56]

The manufacturers of ice machines constituted part of an important but underappreciated set of firms during the height of the industrial revolution: capital equipment manufacturers. Unlike companies that built many units of the same object, these firms built large, unique machines designed to meet the specific needs of individual customers.[57] The sales catalogs of these manufacturers offer no prices because the price of an ice machine always depended upon the specifications that the purchaser required. These manufacturers also kept in touch with their consumers after they installed a machine because ice and cold storage firms often needed help operating it. Since multi-ton ice machines required many months to build and ship to different parts of the country, most of these firms remained small compared with the large-scale factories owned by the Carnegies and Fords of the world. Even the largest refrigerating equipment manufacturers could make only a dozen or so machines each year.

Mechanical refrigeration equipment manufacturers found their first large market among brewers. Before mechanical refrigeration, brewers of lager beer, which required low temperature for fermentation to occur correctly, had to harvest tons of natural ice to make this popular German-style brew. The Lemp

Brewery of St. Louis, for example, harvested its ice on the Mississippi River and stored it in a vast underground cave system near its brewery by the river to keep 50,000 barrels of lager beer cold.[58] The first companies solely devoted to refrigerating equipment manufacturing appeared during the 1870s and early 1880s to meet the demand from brewers, who wanted to find an easier way to make and preserve lager. Most brewers gave up natural ice reluctantly because they feared that their equipment might fail, but they gradually came around as the machines became more reliable and the benefits of this technology became known. One customer for such an installation wrote in an 1883 testimonial, "My cellars and fermenting rooms are kept as cool as I desire, while the air in them is dry and fresh, a marked contrast with the condition when ice is used and which only a brewer can appreciate."[59] Mechanical refrigeration also meant that brewers no longer had to worry about whether they had access to enough natural ice after a warm winter. Mechanical refrigeration also saved them the cost of paying workers to move ice and gave them more precise control over the temperature of their product.[60] Even with such loyal customers, the refrigeration machine industry did not grow quickly until other firms besides breweries started buying their product.

The founder of one refrigeration machine company, John C. De La Vergne, began his career in the grocery business. In 1876, he became a brewer. In the course of trying to produce beer more efficiently, he began to experiment with mechanical refrigeration because of the inconvenience and cost of natural ice. After getting a series of patents, he installed his company's first machine in the Herman Brewery (which was part-owned by De La Vergne) in 1880. Recognizing that there was greater growth potential in selling refrigerating equipment than beer, De La Vergne organized the predecessor of the company that came to bear his name in 1883, just in time to capitalize on increasing concerns with the cleanliness and supply of natural ice.[61] His patents to better seal the valves and lubricate the internal mechanism of the compression pump made his machines better than the other ones on the market during the industry's early years.[62] "That ice-making has at last arrived at a state of perfection," *Frank Leslie's Illustrated Newspaper* reported in 1891, "is largely due to the enterprise and engineering skill of the De La Vergne Refrigerating Machine Company, whose machines have acquired almost a world-wide reputation." By 1891, that firm employed 900 mostly skilled workmen and usually made one massive machine a month, for installations all across America.[63] Despite De La Vergne's unexpected death from Bright's disease in 1896, the firm continued to grow.[64] By 1908, the company had expanded to twice the size of its nearest competitor, and it exported its ice machines around the world.[65]

Perhaps mechanical refrigeration would have spread faster if all ice machines had pleased their customers as much as the De La Vergne Company's machines did. In 1886, the *Ice Trade Journal* declared that many of these other machines were totally worthless except as scrap metal. While they may have created cold, they often proved so inefficient that the fuel costs of operating one surpassed whatever earnings an operator made from ice sales. As long as this technology remained unreliable, expensive repairs also harmed the artificial ice business.[66] Between 1884 and 1890, more than forty different companies began building refrigerating equipment. Few survived to the end of this boom period.[67] "Look at the wrecks scattered along the years of refrigerating history!" lamented the journal *Cold Storage* in 1900. "The majority of those were caused by illegitimate business methods. To sell machines these concerns cut their regular price list rates. Why? Because their machines were not what they claimed to be. By this method these firms rose to sudden prominence, but fell, like the stick, after the rocket exploded."[68] Unless the user lived near the refrigerating equipment manufacturer's factory, the business had to operate the equipment on its own and usually had no recourse when the machinery needed repair either.

Business owners could not even turn to their operating engineers for help, as not even the men who ran mechanical refrigeration equipment knew much about the machines they tended. While the term "refrigerating engineer" eventually came to describe the men who operated mechanical refrigeration equipment in an ice-making or cold storage facility, nobody was trained specifically to run an ice plant during the 1890s. Engineers with general training simply did their best to keep up with this continually changing technology. They usually worked with unfamiliar equipment in isolation and with no quick way to contact the manufacturer if something went wrong.[69] As late as 1900, the engineer Iltyd Redwood could write, "In a great many instances engineers who have charge of these machines only run them by rule-of-thumb methods, and knowing nothing about the why and the wherefore are, in the event of conditions being changed, unable to reason out what will result from the changed conditions, and what other changes ought to be made to counterbalance them."[70] Even if an owner found an expert engineer, that owner might not listen to the engineer if the owner thought he already knew more about the industry than the engineer did.

Luckily for businesses everywhere, refrigeration machinery became increasingly reliable and easier to use as time passed.[71] This boded well for the future of the technology.

Mechanical Refrigeration outside the United States

While the invention of mechanical refrigeration depended upon innovations developed around the world, the result of those efforts never sold as well elsewhere as they did in the United States. In fact, few other countries had anywhere near as many of the giant mechanical ice machines as the United States did because most countries did not have consumer ice industries. In Great Britain and Germany, where large concerns like J. and E. Hall and the Linde Company manufactured mechanical refrigeration equipment, that equipment went almost exclusively for industrial uses. In 1899, Australia had seventeen and New Zealand twenty-two freezing works that used refrigerating machinery for preparing meat for export.[72] Outside of these select industrial circles, hardly anybody even knew about mechanical refrigeration. In smaller countries, mechanical refrigeration hardly existed at all.

A special consular report from the U.S. Department of State entitled *Refrigeration and Food Preservation in Foreign Countries* illustrates how slowly mechanical refrigeration spread throughout the world. The department compiled the report on behalf of the American refrigeration equipment industry in 1890. Basically, the State Department sent a letter to its consuls around the world asking them to describe the state of the refrigeration industry in the places around the world in which they served, and asking what potential American refrigerating equipment might have there. The report published those responses in full. Only a few of the letters sent from nearly 150 consular districts mention the production of any ice at all in their jurisdiction. Many of those responses read like the one from the consul in Turkey, which described two ice factories in Smyrna that ran only six months during the year as the only such facilities in all of Asia Minor.[73]

The few refrigerating machines in service in Great Britain during the late nineteenth century were being used by industry. According to an 1898 issue of the British journal *Ice and Cold Storage*, "Refrigerating machines of various kinds are now extensively used for preserving all kinds of dairy produce; for brewery purposes; fruit importation; bacon curing; India-rubber manufacture; natural-ice skating rinks; preserving fish, poultry and game; chocolate cooling;

gunpowder works; smokeless powder factories; private mansions, hotels and asylums; and last, but by no means least, on mercantile ships of all nations and men-of-war of our own and other Powers."[74] All these industrial consumers explain how Great Britain could have a large mechanical refrigeration industry even though it did not use that technology to make ice for individual consumption. J. and E. Hall sold refrigeration equipment—at first cold air machines, but later carbon dioxide and then ammonia compression systems—worldwide. Between 1878 and 1888 the firm supplied its machines to food importers and dock companies that built and operated cold stores at major ports of entry, rather than for ice plants that operated along the American business model of installing machines to produce ice for home or even industrial use.[75] After it switched over to carbon dioxide refrigeration, the firm primarily supplied machinery for ships that carried food. Without demand among the public, no cold chain existed in Great Britain other than the one for food, preventing ice and refrigeration from having the same impact there that it did in the United States. As late as 1949, a visiting American engineer reported that "in the several thousand miles I traveled in the British Isles, I saw less than a dozen ice manufacturing plants and an ice delivery wagon is a rare sight. Even milk is delivered the year round without the benefit of ice cooling."[76] Argentina, a major exporter of perishable food, produced ice only for "meat-packing houses, breweries, creameries and the like." While many places in the country could have benefited from an ice plant, there was neither domestic nor foreign capital to build one.[77] As a result, no consumer market for ice existed in Argentina during the era before household refrigerators.

Of all the countries in the world, only Germany had a market for mechanical refrigeration that resembled that of the United States in the slightest. "In the manufacture of artificial ice . . . the Germans are masters," wrote the American consul in Frankfurt in response to the U.S. State Department's letter.[78] The fact that Carl Linde perfected and still produced refrigerating equipment in 1890 explains that assessment. Yet the growing pains of the Linde Company demonstrate that not even in the home of mechanical refrigeration did the inhabitants take to mechanical refrigeration in the same way that Americans did. Linde's company maintained the German patent to his own invention, but he opened only four ice works before 1882. The company had sold all the ice works it owned by 1890, preferring to create new companies that would create demand for refrigeration equipment in other areas besides ice manufacturing, such as beet sugar extraction, paraffin extraction, and the liquefaction of chlorine.[79] A 1908 census of the German refrigeration industry found that brewers used 60 percent of the mechanical equipment in the country, the meat industry used 20 percent, 8 percent went for cold storage, and the chemical industry used 7 percent. Only 3 percent of the refrigeration machines in Germany made ice. (In contrast, 40 percent of American refrigeration machines made ice in 1907. Brewers used only 30 percent.) Linde's company exported most of the machines that it made.[80]

In 1893, France still got most of its small supply of natural ice from Norway or from the mountains in the southern part of the country. A single ice factory in Boulogne manufactured "more than the people need," according to the *Ice Trade Journal*, yet fishmongers still used imported natural ice to preserve their catch.[81] This suggests that the Boulogne factory was both technically and economically backwards. Only ten ice plants existed in Japan in 1900.[82] As a result, the consumption of ice did not take off. "It was no uncommon sight to see the Japanese standing in two or three rows of a quarter mile in length waiting their turn to enter" an exposition in Osaka, wrote a correspondent to *Ice and Refrigeration* in 1904, "and it was the only building on the grounds where it was found necessary to station policemen to control the usually placid Jap, so eager was he to see how cold air and frost was artificially produced."[83] Leaving aside the obvious racism here, this correspondent's description of the Japanese reaction suggests the differences between the United States and the rest of the world.

What was mere background noise in the United States remained novel in other parts of the world. As *Cold Storage* observed in 1901, "The feat of artificially freezing water, no matter how hot the weather might be, was once regarded as a great curiosity. It was exhibited in laboratories, as an interesting chemical experiment. The possibilities in store for the invention were only dimly foreseen. But, following the course of all great discoveries, the process was improved year after year, until the present perfection was reached. Now it is within the reach of the populace. The hottest climes are enabled to have what was once an impossibility—pure ice the year round."[84] While the hottest climes could theoretically have ice year round, improving mechanical refrigeration equipment enough to make its product cost effective and efficient remained a serious obstacle. Equally importantly, those technical changes created new uses for refrigeration, which meant that making these improvements was more profitable for the companies trying to perfect the technology.

A Bright Future

Around the turn of the twentieth century, American refrigerating equipment manufacturers began to systemize their production process. The efficiencies this created led to a decrease in prices that greatly benefited the industry and the public at large. What once got done in an improvised manner increasingly was based on precedent and past practice. As American engineers gained hands-on experience with mechanical refrigeration, the scientific principles behind artificial cold became clearer, and the reliability of refrigerating machinery improved.[85] Small, incremental improvements in all three components to the refrigeration system—the compressor, the condenser, and the evaporator—improved the efficiency and reliability of every mechanical refrigeration system. After 1890, automatic control of the flow of refrigerant as well of the overall operation of the machine became possible for the first time. This led to the development of smaller machines. The gradual shift from steam to electricity to power all kinds of refrigeration systems, especially the smaller ones, played a particularly important role in cutting costs. With power generated outside the plant, operators needed significantly less machinery inside the plant. This meant that fewer things could go wrong.[86]

All these changes led to a significant improvement in the quality of the ice that mechanical refrigerating equipment produced. As late as 1893, the *Ice Trade Journal* suggested that artificial ice "lacked staying power," which meant that it melted too quickly when stored.[87] While in previous decades, the product of mechanical ice machines "was little better than snow," wrote one unnamed observer of the industry in 1897, "within the last few years there has been a marked reduction in the cost of these machines. . . . Such a machine to-day costs less than half of what it cost five years ago."[88] These kinds of improvements also reduced the time needed to make a ton of ice from hours to just minutes.[89] This too helps explain why increasing numbers of customers switched from natural to machine-made ice. None of these improvements came as a result of better understanding of underlying principles related to refrigeration. They came through better adherence to the science that refrigerating engineers already knew.[90]

As efficiency improved, the size of refrigerating machinery shrank. This was crucial for the development of the modern cold chain, as it eventually made it possible for mechanical refrigeration to reach small, inconvenient, and sometimes mobile places like kitchens or railway cars. The advent of these smaller, more efficient machines made mechanical refrigeration much more popular than back in its early days. Before mechanical household refrigeration became feasible during the 1920s, only businesses or very rich households could manufacture their own ice. Now small manufacturers, storekeepers, hoteliers—anyone with only a moderate demand for ice—were able to make it themselves if they invested in the necessary equipment. In 1900, the *New York Tribune* reported that "hundreds of hotel keepers, butchers, saloonkeepers, dairy and delicatessen store keepers, apartment house proprietors and owners of big private houses in this city" had taken this course of action to save the expense of buying ice.[91] In 1907, as this trend grew, the *Cold Storage and Ice Trade Journal* noted that such machines were "the results of many years of the hardest kind of study and endless experiments and the outlay of a large amount of money."[92] This development also had the benefit of better matching supply and demand, since small businesses could make exactly as much as they needed. Most ice consumers still could not afford the capital outlay to buy and run their own ice machines. Because of this situation, a large demand for new ice plants that would satisfy the household market persisted after the turn of the twentieth century.

Despite little initial consensus over exactly what the job of refrigerating engineer involved, refrigerating equipment manufacturers and other industry representatives dispersed information on the best ways to operate different kinds of refrigeration equipment, so that over time they became common practice. The founding of *Ice and Refrigeration* in 1891 marked an important milestone in the distribution of such information.[93] The best practices that this journal publicized helped minimize the differences in efficiency between different kinds of refrigerating machine equipment. By the end of the first decade of the twentieth century, the results obtained with any of the types and makes of machines on the market became essentially the same. "Each of the representatives of the various machines make claims and statements which would seem to indicate that his machine is superior to all others," wrote two operators in an article published in the journal *Electrical World* in 1908. "Our experience indicates that there is very little actual difference in results obtained from the different makes of machines, but that the manner in which the equipment is planned and installed has much to do with the giving of good or poor results."[94] This proved true of machines that operated with different refrigeration systems (carbon dioxide versus ammonia compression, for example) as well as of different machines that used the same setup to produce cold. The mechanical ice machine had come of age.

Inside and outside the industry, people assumed that the technology of mechanical refrigeration had a bright future. "It may safely be assumed," wrote the author of one important reference book about this technology in the early twentieth century, "that the comparatively new science which forms the subject of this treatise is one that will develop into a special industry of vast proportions that will have no end, until the present organized constitution of the globe is essentially changed." This engineer presented a uniformly positive picture, when, in the "not far distant" future, "men will wonder that the world was ever able to exist without its aid."[95] While observers might have expected this attitude from an engineer, the reception that mechanical refrigeration received from the general public indicates that this attitude had at least some basis in truth. The small businesses that bought refrigerating equipment for the cold chain did not consume ice itself. For them, ice machines and cold storage units served as a means to an end. The consumers of the food transported along the cold chain did not care how their food got preserved as long as the technology preserved it effectively.

Entrepreneurs ordered expensive refrigerating equipment long before the technology itself became completely reliable because they saw the great demand for ice among the public. Potential ice machine operators wanted so much of this machinery that during the late 1880s and early 1890s that some equipment manufacturing firms with good technology at their disposal went bankrupt trying to fulfill contracts for machines that they had guaranteed in advance.[96] An "open winter" in 1889–1890—one too warm for natural ice to form—gave a significant push to the building of ice machines even in northern cities that had traditionally depended upon natural ice. The number of ice plants in America increased from 222 in 1890 to 1,320 in 1905. The total capacity of mechanical refrigeration in America increased from 186,590 tons per day in 1904 to 284,780 in 1909. In 1890, the Bureau of the Census recorded only 222 ice plants in the entire United States.[97] By 1907, the Census Bureau could report that manufactured ice had "passed beyond the experimental stage and is now firmly established."[98]

As the mechanical refrigeration industry grew, people with no experience with ice and refrigeration became increasingly comfortable with the idea of operating this machinery. Thanks to technological improvements over the previous twenty years, operating a mechanically driven ice plant did become much easier. No longer inhibited by the inability of engineers to operate its machinery, American industries in impressive numbers came to depend upon mechanical refrigeration by the turn of the twentieth century. A writer for the journal

Power in 1910 concluded that there were a total of 143 different uses for refrigerating equipment besides making ice or just keeping things cold.[99] As a result of such new uses, many of the firms that bought refrigerating machines were not primarily ice providers. Advertisements in old copies of *Ice and Refrigeration* reveal other new applications for this technology as they developed. Besides stabilizing dead bodies in the morgue, the journal reported on refrigerating molten steel during steel manufacturing and refrigerating flowers at the florists (just to name a few uses). As in Britain, mechanical refrigeration also became a fixture in industrial processes such as refining oil products, separating gaseous mixtures, treating textiles with lye at low temperatures during their manufacture, and producing rubber, pharmaceuticals, and even explosives.[100] While none of these uses were part of the cold chain, the spread of refrigeration made consumers and producers increasingly comfortable with the idea of using it for preserving and transporting food.

Power companies constituted a particularly important group of new buyers for refrigerating equipment during these years. They entered the ice business in a big way during the first decade of the twentieth century because making ice allowed them to utilize all the surplus power they generated before most households were wired for electricity. By 1910, 122 power companies made by-product ice, ice manufactured using this surplus power.[101] Ice season came at a perfect time for utilities, since demand for electricity was lowest during the summer in the days before air conditioning. Demand increased again as the weather cooled, precisely when the ice business slacked off.[102] In many instances, utilities made more money selling ice than they did selling power. As these operations grew, it became apparent to countless power operators unfamiliar with the ice business that even a poorly run ice-making division could turn a profit, since power was the most expensive cost in producing ice and these utilities had power to spare. According to the journal *Electrical World*, "Where ice making is conducted as a separate enterprise it is almost always a money-making proposition." For this reason the journal recommended that many power station owners consider abandoning the sale of electricity altogether and just sell ice.[103] Even though power companies did not entirely understand how their refrigerating equipment worked, they could still make money in the burgeoning ice industry.

Since mechanical ice machines produced a steady supply of product no matter what the temperature had been the previous winter or the state of competition in a particular market, refrigeration technology did much to smooth out the boom or bust cycles of a theretofore unstable industry. Nothing except the price of fuel caused artificial ice companies to seriously fret. The industry became established as the scale of costs and benefits that an operator weighed tipped from one side to another. Ice companies, not households, bought mechanical refrigeration at this stage of the technology's development. From the standpoint of individual consumers, ice was ice whether made by machine or cut from a pond. Machine-made and natural ice may have looked somewhat different, but inside an icebox just the temperature of the food really mattered. Only after about 1890 did competition between ice firms lead consumers to differentiate between the machine-made and the natural product. At points along the cold chain that the consumer did not see, the manner of refrigeration did not matter so long as the food that needed preservation stayed cold.

As refrigeration technology changed, the nature of the cold chain changed too. Between 1880 and 1910, the erection of numerous local ice manufacturing plants shortened cold chains throughout the country, diminishing the importation of natural ice or sometimes making it totally unnecessary. With improvements in mechanical refrigeration technology, many communities in the central United States and more than a few towns in the North became sites of fierce competition between natural and artificial ice providers. This became a fight to the death for these two closely related industries. While many observers successfully predicted the eventual winner, the fighting lasted decades because even the overmatched combatant had some advantages that its new opponent did not. Only when that struggle ended could the cold chain take its modern form.

CHAPTER 3

The Decline of the Natural Ice Industry

In the late nineteenth century, the Schuylkill River, the main source for Philadelphia's drinking water, became infamous for its pollution.[1] As early as 1875, the chemical engineer Julius W. Adams, a consultant for the city, reported that the Fairmount Pool, the reservoir into which the Schuylkill drained inside the city limits, "is, at times, from the amount of refuse, from the slaughterhouses, breweries, and above all, the manufactories at Manayunk, not a proper water for domestic use. This is conceded by all who have examined it."[2] Smokestacks became symbols of progress during the nineteenth century, but Philadelphians knew they had a problem when chemicals entered their source for drinking water. They not only drank that water every day; they also consumed ice cut every winter from that same river when summer came around. Every glass of water chilled with natural ice left sediment as the ice melted, but as the pollution worsened, consumers increasingly recognized that industrial chemicals already contaminated their ice.

While the city of Philadelphia had banned cutting ice from the Fairmount Pool around this time (presumably for sanitary reasons), natural ice companies cut thousands of tons of ice from sites further up the Schuylkill and on its tributaries during the late 1870s and early 1880s despite the pollution there.[3] Ice firms did not think that consuming this ice carried any health risk because they, as well as many independent experts, incorrectly believed that freezing was self-purifying—in effect, that disease germs trapped in ice froze to death. (In fact, they starve.) In the winter of 1882–83 the stench of chemicals from the Schuylkill became so bad that even Philadelphians long accustomed to polluted water took notice. One paper described the Schuylkill as having "a dark greenish color," with "thousands of dead fish . . . lying about the banks."[4] When the water flowed, Philadelphians saw woolen and cotton fabrics float by; large amounts regularly appeared in the water north of Vine Street, the part of the city closest to the river from which polluted ice would have come.[5] Ice manufacturers claimed that the freezing process expelled all such impurities. The city's water department suggested that when the ice harvesting firms cut the ice on the river open, the impurities would waft away. The Knickerbocker Ice Company, the largest in Philadelphia, cut the ice, tested it free of charge, and pronounced it safe.[6] The people of Philadelphia thought otherwise.

Knickerbocker's Philadelphia ice cards from these years indicate the extent to which consumers rejected their local product. Ice companies distributed these cards to display in windows on days when customers wanted ice delivered to their homes. The cards included not just a company logo, but also price information for a particular year as well as information about the product. The Knickerbocker Ice Company's card from summer 1882, before the river water became unbearable, includes no language about the origins of the ice. But the card from summer 1883 has the words "We furnish PURE EASTERN ICE ONLY" stamped on it, suggesting that the company had come to realize how the market had changed in just one season. Eastern ice might be a euphemism for ice from Maine, but the important point was that Knickerbocker's ice did not come from the Schuylkill. By 1884, the words "PURE IMPORTED ICE" were printed on the card in bold. The similarity of the cards in every other respect across this time frame only underscores the significance of this change. This large, well-established natural ice supplier had to adapt or die. It did so by finding new sources for its ice.

As natural ice grew dirtier, mechanical refrigeration gradually became the primary technology along the cold chain. Pollution alone does not explain this change. Consumers could buy natural ice for only part of the year, and even in winter its quality depended upon the weather. Ice machines could produce all year round. Natural ice took up lots of space and usually required a long trip to get from its point of production to its point of consumption. Artificial ice also took up lots of space, but machines produced it in regular shapes. This made it easier to transport, maintain, and control with precision. Because artificial ice surpassed natural ice in terms of quality in every way, it became a vital element in the development of the modern cold chain. Only with the development of small mechanical refrigerating units capable of fitting into every link of that chain did natural ice completely disappear.

The only advantage that the natural ice industry had in its battle for survival was its product's low price compared with that of artificial ice. This allowed the natural ice industry to thrive for decades despite the numerous disadvantages of its product and even in the face of stiff competition. The prolonged willingness of many Americans to consume a dirty and inconvenient product like natural

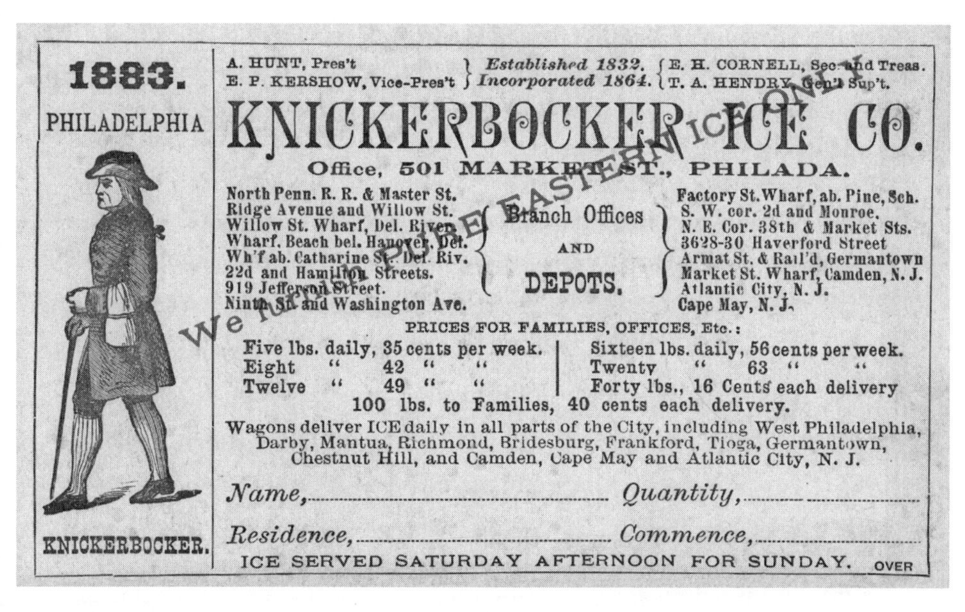
1883.
PHILADELPHIA
KNICKERBOCKER.
A. HUNT, Pres't
E. P. KERSHOW, Vice-Pres't
Established 1832.
Incorporated 1864.
E. H. CORNELL, Sec. and Treas.
T. A. HENDRY, Gen'l Sup't.
KNICKERBOCKER ICE CO.
Office, 501 MARKET ST., PHILADA.
North Penn. R. R. & Master St.
Ridge Avenue and Willow St.
Willow St. Wharf, Del. River.
Wharf, Beach bel. Hanover, Del.
Wh'f ab. Catharine St., Del. Riv.
22d and Hamilton Streets.
919 Jefferson Street.
Ninth St. and Washington Ave.
Branch Offices AND DEPOTS.
Factory St. Wharf, ab. Pine, Sch.
S. W. cor. 2d and Monroe.
N. E. Cor. 38th & Market Sts.
3628-30 Haverford Street
Armat St. & Rail'd, Germantown
Market St. Wharf, Camden, N. J.
Atlantic City, N. J.
Cape May, N. J.
We furnish PURE EASTERN ICE ONLY
PRICES FOR FAMILIES, OFFICES, Etc.:
Five lbs. daily, 35 cents per week.
Eight " 42 " "
Twelve " 49 " "
Sixteen lbs. daily, 56 cents per week.
Twenty " 63 " "
Forty lbs., 16 Cents each delivery
100 lbs. to Families, 40 cents each delivery.
Wagons deliver ICE daily in all parts of the City, including West Philadelphia, Darby, Mantua, Richmond, Bridesburg, Frankford, Tioga, Germantown, Chestnut Hill, and Camden, Cape May and Atlantic City, N. J.
Name, ... *Quantity,* ....
Residence, ... *Commence,* ....
ICE SERVED SATURDAY AFTERNOON FOR SUNDAY. OVER

Customers displayed ice cards like this one in their windows on days when they wanted ice delivered. The words "We furnish PURE EASTERN ICE ONLY" stamped on this one suggest the unpopularity of ice from Schuylkill River the year after the river had a serious pollution problem. Courtesy of the Hagley Museum and Library.

ice demonstrates how much they valued the ability to control the cold. Even poor urban immigrants spent precious pennies on ice to keep cool during long, hot summers. The other major market for natural ice, industrial consumers, did not even care about its cleanliness. They only wanted the cold it provided. This explains how even a quintessential nineteenth-century technology like this one outlived its era. Mechanical refrigeration had to become cheaper and more convenient to compete with inexpensive natural ice.

The Cleanliness of Natural Ice

In the last two decades of the nineteenth century, people in cities across America turned to iced drinks to cool down. Every consumer of these beverages expected that dirt and other sediment in the natural ice would remain at the bottom of the glass. "Natural ice contains, always, a certain amount of impurities, gaseous and solid, no matter where it is obtained from," explained a manufacturer of mechanical ice-making equipment in 1899. "These impurities are imprisoned in the ice and released only when the ice is melting."[7] Particles trapped in ice were "by no means so infrequent as to warrant any surprise when . . . encountered," remarked the *New York Times* in 1901.[8] Natural ice left dirty glasses, but the benefits of cold drinks over tepid ones outweighed whatever concerns consumers might have had about cleanliness.

Besides whatever got frozen in with the water, the manner in which the harvester shipped also resulted in sediment inside even the best natural ice. Often producers shipped natural ice between straw or wood shavings. Some of that inevitably rubbed off onto the product. Similarly, when harvesters cut holes in the ice to let water flow to the top so that it could freeze more evenly, that water would capture all the impurities on its surface. Those impurities remained until the ice melted, often leaving a foul muck at the bottom of people's iceboxes or a black scum at the bottom of a glass of water. During mild winters when the ice did not freeze to a sufficient depth, harvesters would stack two thin sheets of ice to make a single thick one. Ice trapped in the middle of these two halves had the same unfortunate trait.[9]

A market existed for even the dirtiest ice, since it still met industrial needs as long as it did not come into direct contact with foodstuffs. Two such uses were in grocers' iceboxes and in taverns. The commercial iceboxes of this era had separate compartments for the meat or produce displayed in them and the ice used to keep them cold, specifically to avoid the possibility of contamination. Similarly, polluted ice could keep beer barrels cold both before and after tapping. Unfortunately, unscrupulous dealers often substituted dirty ice for the clean kind for general consumption, since cleaner ice cost more. Nobody could examine the entire surface of large quantities of ice before delivery, and since even clean ice had some dirt on it, buyers had trouble telling the difference between different grades of ice even after it melted.

Despite some sediment, ice harvested properly from a clean body of water could yield a product that satisfied even the most fastidious consumers. For example, Henry David Thoreau's Walden Pond attracted ice cutters during his famous stay on its shores because of its deep, clear blue water. This created clear ice, which meant that cold drinks with Walden Pond ice tasted less of sediment and that consumers did not have to avoid that sediment as the ice melted. Thoreau, a kind of self-taught expert on water quality, described the natural sediment in Walden Pond's ice: "Like the water, the Walden ice, seen near at hand, has a green tint, but at a distance is beautifully blue, and you can easily tell it from the white ice of the river, or the merely greenish ice of some ponds, a quarter of a mile off."[10] That greenish tint came from the organic matter in the water.

Thoreau found Walden Pond's ice beautiful—a mere touch of green rather than greenish—because the pond's deep waters diluted the organic materials. As the *Ice Trade Journal* explained in 1881, "It is in accordance with the experiences of all intelligent dealers, that the depth of the water in any given place has much to do with the color, and often with the other qualities of the ice formed on its surface, and that, as a rule, the deeper the water the better will be the product. It will not make so rapidly, but it will be purer and more transparent."[11] Unfortunately such places became increasingly hard to find, both because of pollution and because of the fierce competition among ice harvesters to find such bodies of water within easy reach of America's cities.

The best natural ice came from streams because the motion in the water both promoted the growth of ice and made it more difficult for sediment to get trapped in it. A gentle current in a pond or stream had the same effect. The current also helped air escape, an advantage for harvesters, as it prevented air bubbles in their product. This also kept the ice clear, and consumers invariably liked clear ice because it seemed purer than the cloudy kind. Ironically, air bubbles can form in even the purest ice, so transparency did not indicate quality. To cater to the preference for clear ice, harvesters sometimes cut channels to release water from ponds in order to promote a current strong enough to clarify the ice but not strong enough to retard its growth.[12] The worst natural ice always came from the top of a solid body of water because sediment tended to get trapped there during the freezing process.[13] Ice cut from the top of a body of water sold at a lower price than better ice cut from the bottom, since consumers could easily see the sediment trapped in it. Artificial ice dealers made the most of the variable quality of the competition's product. "Don't take the ice man's word for anything," warned one artificial ice delivery firm in 1914. "Are your family and children in danger?" it asked under a picture of a skull. "We do not handle lake ice."[14] Over time, delivery firms that bought their product from multiple suppliers had to distinguish between natural and artificial ice because their customers grew to prefer the latter, which they saw as the safer product.

As water pollution grew worse not just in Philadelphia but also elsewhere in America, consumers concerned about the safety of ice made with tainted water demanded a product harvested from cleaner sources or they began using manufactured ice. Around Chicago, for example, ice dealers moved from local sources to the still-pure waters of the Fox River up in Wisconsin.[15] In Detroit, customers switched from natural to manufactured ice around the turn of the twentieth century as the pollution by gas companies of rivers in that area grew worse.[16]

One firm in Massachusetts, forced to move its operations to New Hampshire, bought several hundred acres of land around the lake where it cut ice to prevent runoff from other uses of the land and ensure the purity of its ice.[17] Even though the rivers in other cities did not stink or have wool floating in it like the Schuylkill did, companies found better ice sources anyway because their customers demanded it.

To make matters worse, natural ice was often brittle and harvested in uneven blocks. Only a saw could cut through it. Manufactured ice, on the other hand, came in regular-shaped blocks. Anyone could easily cut those blocks with a pick to fit them inside any size icebox.[18] (This also meant that manufactured ice stacked more easily, making it easier to transport in wagons and, later, trucks.) The inviting appearance of manufactured ice came as a result of a deliberate effort by its producers. As transparency became widely associated in the public mind with quality and purity, manufacturers went to great pains to eliminate the harmless air from their product since it would freeze into opaque white areas of ice blocks. Stirring water before freezing it produces clear manufactured ice—any kind of agitation worked the same way that a current would on a stream. But this meant that the stirrer had to be removed before the water froze, thereby creating an extra step in the manufacturing process and possibly increasing the price. Manufacturers could also produce clear ice by keeping air out of the water before it went into whatever mold shaped the ice block. They did this either by putting the water in a vacuum or by boiling it for an extended period of time.[19] These steps did not change what ice did for consumers. It just made them more likely to buy it.

Ice manufacturers often touted the transparency of their product in promotional literature that showed people behind blocks of transparent ice or objects trapped within it. Mark Twain visited an ice plant in New Orleans that froze flowers, dolls "and other pretty objects" into some blocks for effect since they "could be seen as through plate glass."[20] Once the public associated transparency with quality, transparent ice generally cost more than opaque ice no matter how it was made. Presumably the strong public demand for clear ice at a higher price justified these additional production costs. As consumers came to realize that the quality of ice did not depend entirely upon its appearance, additional steps designed to make good ice look better became unnecessary, particularly as such steps inevitably drove up the cost of ice production. Of course, nobody ever worried about the appearance of ice destined to serve as part of the cold chain as opposed to the product sold to consumers.

Because of its clarity, artificial ice seemed more natural to the public than the product that actually came from nature. And since natural ice could have contaminants, artificial ice actually was safer than its competitor. Absent industrial uses, pollution might have destroyed the natural ice industry much earlier than it did. Cutters had already begun to abandon previously lucrative ice fields around the country for this reason as early as 1891.[21] Once artificial ice became serious competition, ice harvesters began to take more care with their product. Reform efforts, even unsuccessful ones, attempted during the Progressive Era helped inspire those changes. Read the pages of old industry trade journals, and it is easy to find quotes from state and local politicians who proposed regulations to protect the health of natural ice consumers. Even the federal government eventually got involved to a limited extent. The Pure Food and Drug Act of 1906 provided the legal basis for protecting consumers from adulterated food, but it did not help here. Nobody considered natural ice taken from polluted waters adulterated until the attorney general of the United States enforced it against the Washington, D.C., subsidiary of the American Ice Company in 1910. In that case, the $150 fine the firm paid as a result could not have served as much of a deterrent.[22] Other public health officials periodically tried to end harvesting from polluted waterways, but these moves changed little, since no visible indicators of pollution existed. After all, most natural ice contained at least some sediment.[23] Some of that sediment was just a nuisance. Some of it was much worse. Nonetheless, the primary risk that sediment of any kind presented came not from germs, but from how the possibility of danger affected the reputation of the product.

Typhoid Fever

Over the course of the 1880s, consumers became increasingly concerned about ice contaminated with pathogens that they could not see. When poor people got stomachaches or something far more serious from consuming natural ice, how would they trace their ailment back to the ice? How would they even know that they had a disease caused by something they ate? As the *Detroit Tribune* explained in 1903, "It is probable that many of the summer ailments of the digestive tract are due to the use of villainously impure ice, which is loaded with dormant bacteria."[24] Nobody will ever know how many people actually got sick from consuming ice. Over time, the scientific community gradually formed a consensus about the dangers of ice, despite assurances by the natural ice industry that ice purified itself through freezing.

For people who did not read the scientific literature, a series of well-publicized outbreaks of typhoid fever traced to natural ice did more than anything else to turn them against this product. The most important of these incidents occurred during the winter of 1901–1902, when the superintendent of the St. Lawrence State Hospital, a mental institution in Ogdensburg, New York, authorized opening a new part of the nearby river to cut ice for the patients and staff. Unfortunately, that part of the river lay only a few hundred yards from the hospital's main sewer line. Investigators later traced a typhoid fever epidemic among the patients and staff that resulted in several deaths directly back to that ice. Advocates of manufactured ice pounced on the opportunity this tragedy presented. "Why on earth should the already enfeebled health of a hospital patient be further jeopardized by the use of microbe polluted ice," asked the journal *Cold Storage*, "when such a comparatively simple matter to erect an ice factory [could] supply an absolutely pure product?"[25] Many people had this kind of concern about the cleanliness of the natural product, but this outbreak made for particularly effective propaganda because the hospital's relative isolation made this the first time that investigators could trace a typhoid outbreak back to ice rather than the water from which that ice came. This incident confirmed the conclusion that typhoid fever bacteria could survive freezing in the minds of the general public for the first time.[26]

Typhoid fever had threatened humankind for centuries. It is caused by a microorganism that attacks the intestinal tract. This microorganism gets carried by human feces and spread through poor sanitation. Its initial symptoms include fever, dehydration, diarrhea, abdominal pain, chills, headache, cough, weakness, and sore throat. Because of the similarity of these symptoms to those of many other diseases and because it takes from between five to twenty-one days for symptoms to appear, typhoid fever often gets misdiagnosed, even today. (Rashes with red spots always reveal typhoid infections, but these spots appear only in 30 percent of cases.)[27] In 1910, the general death rate from typhoid fever was 17.8 people per 1,000, with most of those fatalities concentrated in cities with fouled water supplies.[28] Undoubtedly, many more suffered from this disease without authorities' recognizing that they had it.

While many people suspected the relationship between lack of sanitation and typhoid fever, nobody could prove this link until the German bacteriologist Karl Eberth established a direct connection between feces and this disease by discovering the typhoid fever bacillus in 1880.[29] At this point, with the germ theory of disease still in its infancy, many doubters remained. Scientists had particular trouble finding typhoid bacilli in a sample of drinking water. Unable to see the definitive proof, many scientists still doubted the connection between typhoid and poor water quality (let alone ice) because nobody found these microbes in water until years later.[30] Unfortunately, even a tiny quantity of typhoid microbes could do enormous damage. Some people act as carriers of the disease without showing symptoms (think of "Typhoid Mary" Mallon, a cook by trade). Yet if the waste of even one such individual made it into the water supply, it had the potential to infect thousands of people.

The first suggestion that consuming infected ice posed a risk for contracting typhoid (as well as other waterborne diseases) came from scientific tests of ice samples in the 1880s and 1890s. "Modern researches into the origin of disease leave no doubt that impure ice is frequently the cause of typhoid fever," *Frank Leslie's Illustrated Newspaper* reported as early as 1882.[31] Dr. William Blackwood, writing in an 1893 medical bulletin, explained: "Most folks think that ice 'purifies' itself in freezing by some mysterious process, but how it does this they can't say." On the contrary, argued Blackwood, "It is very well known that many germs which are inimical to health are not killed by even long continued cold, even if the temperature is carried far below zero. Among these, it is said that the bacillus of typhoid fever is highly resistant to freezing."[32] The 1902 outbreak in Ogdensburg led to a flurry of scare reporting in the New York press about the dangers of natural ice cut from the Hudson River, which New Yorkers still consumed by the ton every summer. "We have entered upon the icewater season," editorialized the *New York Times*, "and perhaps the best thing to do is to give thanks that this particular sin is not as summarily or invariably punished as, theoretically, it should be."[33] While the natural ice industry resisted the connection between its product and sin, its leaders no doubt gave thanks every year that public health authorities had not linked natural ice to anything worse than isolated outbreaks of disease.

The public, unable to see the beginning of the cold chain that brought their ice to them, gradually shied away from natural ice anyway. Growing popular sentiment that natural ice posed a threat to the public health endangered the viability of this industry. Confronted by overwhelming evidence that typhoid germs could survive freezing, industry representatives began to argue that they could not survive long enough in an immobile state to infect consumers.[34] Unfortunately, the industry's self-serving interpretations of how the science worked proved inaccurate. Some bacteria did live long enough to infect people, particularly with a shorter cold chain and the development of a market for ice during the winter months.[35] While modern Americans tend to think of things that are "natural" as being superior to the man-made, when it came to ice the opposite proved true. The natural qualities of ice made people worry about their safety when they consumed it. Natural things, like sediment, gave the ice industry problems because nobody wanted to ingest dirt or leaves, or even just clean them out of their iceboxes.

Despite their disadvantageous position in this argument, natural ice producers did not stand idly by and let artificial ice producers take full advantage of this issue. They and their allies often tried to claim the "natural" high ground. "The difference between manufactured ice and natural ice is that one is a live product of the Creator, while the other is the dead product of mere man and his little machine," wrote a *Chicago Chronicle* reporter in 1903.[36] Natural ice firms continued to tout the safety of their product long after scientists had firmly established the possibility that natural ice could convey disease.[37] Even when natural ice looked clean and wholesome, persistent newspaper stories about microscopic germs told a different story. By the time the natural ice industry began to make serious efforts to improve the appearance of its product, that product's reputation with the public had already slipped irrevocably in comparison with the machine-made article. Luckily for the industry, natural ice had a different selling point other than safety that kept it from disappearing for a considerable period of time.

The Price of Ice

Upton Sinclair's extensively researched muckraking classic *The Jungle* suggests that one of the ice dealers in Chicago used to cut at least some of its product from the stagnant pools around Packingtown and sell it to the residents of the neighborhood. Why would anybody buy it? "They did not read the newspapers," Sinclair writes, so "their heads were not full of troublesome thoughts about 'germs.' "[38] Perhaps the poor immigrants who worked in the meatpacking plants of Chicago might have preferred better ice simply on aesthetic grounds, but they did not make enough money to worry about how their ice looked. If they purchased any ice, it had to be the cheapest ice possible. From the time when artificial ice first became a viable commercial commodity through the moment when the natural ice industry finally disappeared, natural ice generally cost less than the machine-made product. Since these two products performed the same functions, that price difference persisted no matter what the weather or market conditions for ice in general. After all, had natural ice dealers tried to set the price of their product at the same level that artificial ice dealers did, consumers would likely have opted for the competition on aesthetic grounds alone. The price discount made up for natural ice's other risks and shortcomings. Unfortunately for the natural ice industry, that advantage could only keep it afloat for so long.

For as long as natural ice and artificial ice competed against one another, the prices generally remained unstable for both products.[39] The price of natural ice was particularly unstable because of that industry's dependence upon the weather. The price of natural ice became a matter of controversy, as the *Philadelphia North American* explained in 1893, because "the causes controlling the market price of ice are something far beyond the understanding of the heated public."[40] Most consumers in northern cities expected cheap ice. After all, ice surrounded them every winter: Why should it cost any more this year than last? In fact, since it did not cost natural ice companies anything to make, why should it cost anything at all? Many companies felt the need to answer such questions publicly in order keep demand up. As a result, they began a conversation in the media with ice consumers. This conversation illustrates precisely why looking at price as simply the interaction of supply and demand does not tell the whole story of how transactions for ice happened. Since it appeared that the good had no value, consumers required proof otherwise. Ice firms had to convince buyers that they sold more than just transportation and service.[41]

Since ice markets had become entirely local by the late nineteenth century, nobody ever compiled detailed information about ice prices.[42] Once firms like Tudor's stopped shipping ice to New Orleans or Calcutta, almost nobody shipped it by water anywhere, and it never became an important cargo on railroad cars in its own right. With the exception of Maine, natural ice tended to get consumed near the location where it was harvested. As the length of the cold chain for each kind of ice shrank, the cost of production (rather than the cost of transportation or storage) became the most important factor in determining the relative prices of each good. Between 1890 and 1910, natural ice generally cost less in cities where natural and artificial ice competed because the former required little expensive machinery and used more horsepower than expensive fuel. "In the present state of ice manufacturing processes there is said to be no possibility of being able to make and sell artificial ice as much less than thirty cents per hundred pounds delivered," explained a Massachusetts ice manufacturer in 1906, "while natural ice, after a severe winter, can be sold at a much lower price."[43] Similarly, a representative of the Passaic Ice Company, an artificial ice firm in New Jersey, complained in 1899: "About 75,000 tons have been harvested this winter, about twice as much as last winter. If prices will not be cut too much by big outside companies the prices may be kept at $1.50 to $2 per ton, that is $1 less than last season. It is hard for artificial ice companies to compete with such prices."[44] Natural ice dealers could survive an intense competition with ice made by the increasingly efficient mechanical alternative as long as they could continue to undersell it.

The difference between the prices of these two kinds of ice generally stayed fixed, but ice prices in general fluctuated wildly, depending on any number of factors. Of course, the weather had the greatest impact on the price of natural ice from one year to the next. Warm weather in the winter hurt supply because whatever ice formed was thinner. "It is highly injurious to have ice freeze solid at night, then melt a little under the rays of the sun, and then freeze again at night," explained a New York City dealer in 1894. "It honeycombs it and makes it unfit for storage."[45] The quality of natural ice also suffered if the very cold days in winter did not last particularly long. Similarly, hot weather in the summer increased demand and often led to increased prices. Ice prices also rose for many other reasons. Sometimes ice companies tried to pass on the high costs for material and labor to consumers. In other instances, lack of competition allowed firms to test higher price points on their customers. If an ice company bought from an ice harvesting firm instead of harvesting its own product, it might raise prices when its suppliers raised prices.

Ever-growing consumer demand also served as a common justification for high ice prices during this era. As the market expanded, it became increasingly difficult for companies to predict how much ice they needed any given year. In particularly hot weather, ice dealers often ran out—an event dubbed an ice famine. The industry often justified raising prices as a way to head off such an event. Sometimes companies induced ice famines on purpose. "The large companies have plenty of ice, enough to stand the present demand as large as it is," reported a smaller ice merchant to the *Brooklyn Daily Eagle* in 1876. "Because some of the smaller companies are pinched, and compelled to purchase where they can supply their customers, a general advance in price is agreed upon."[46] This split in the ranks reflects the structure of the industry. Some companies cut and supplied ice. Smaller merchants bought it from wholesalers, so these dealers both bought and sold ice. Any of these problems could lead ice companies to raise prices, which they did on a regular basis. An increase in the price of natural ice often caused artificial ice prices to increase as well, since their product filled the same market niche and the more affluent customers for that product could afford to pay a premium for what most people perceived to be a superior product.

Ice prices did, of course, sometimes go down. For example, a long, cold winter could lead to an excess in the supply of the natural product and therefore lower prices all around. Less obviously, if any dealer had a lot of ice on hand melting in storage, he always had the option of dumping it on the market at a significantly lower price so as to recoup at least some of the production costs. To make matters worse, the less you had, the faster it would melt. This often led to panic selling by dealers desperate to earn all they could from their remaining inventory. Unfortunately for their colleagues, price reductions by one dealer tended to be matched by others fearful of being put in the same position. Keeping ice prices low required an effective distribution system. Ice might cost much more in towns just a hundred miles outside of a large city if no wholesale distribution system existed for people to purchase it there.

When customers complained about ice prices, both natural and artificial ice dealers usually responded by pleading poverty. While this kind of whining may seem like what any smart businessman would do, ice companies of both types did operate under very thin profit margins. Anthony Vizard, a New Orleans dealer, told the *Daily Picayune* in 1890: "No [ice company] has ever been known to pay a dividend, and many have been obliged to mortgage their works to secure money to keep running. You can go on the market now and buy the shares of any company for 50 [cents] on the dollar."[47] Because of their small margins, ice companies competed fiercely, especially in large cities with large markets for ice like New Orleans. Other ice dealers responded to criticism about their prices by getting into greater detail with regard to the costs associated with their business. "The loss from melting during all this transportation is never less than 20 percent in hot weather," one harvester wrote in anonymous letter published by *Ice and Refrigeration* in 1898, "and will sometimes run up as high as 50 per cent. The freight charges on the railroads are also a large bill of expense. Taking all these things into consideration, the cost of natural ice is climbing upward every year, and the natural ice companies see their profits reduced as the expenses increase."[48] These kinds of costs did not concern artificial icemakers, since they tended to operate much shorter cold chains. But the natural ice industry's entire business model teetered on the brink of collapse when their comparatively low equipment cost was offset by higher costs from storage and transportation.

To end brutal price competition, firms in the industry tended to consolidate over time. As an added benefit, greater size also led to efficiencies in storage, transportation, and production. During the 1880s, consolidation usually took the form of local cartels. Consumers generally did not approve. "Having the idea that 'the Lord made a big crop of ice' last winter," explained *Ice and Refrigeration* in 1893, "they consider, therefore, that the price this summer should be merely nominal; and whenever dealers meet together and agree to stick to a fixed scale, immediately the press sets up to wail."[49] From consolidation, companies took the next logical step and formed local monopolies. In Chicago, twenty-seven ice companies came together as a single firm in 1898. It controlled not only the ice business there but also the most important lakes in Wisconsin and northern Illinois.[50] In St. Louis, seven firms formed the Polar Wave Ice and Fuel Company shortly after 1900. Larger and smaller firms met their colleagues at regional and national ice association meetings, where they discussed the problems that faced the industry, such as legislation regulating the purity of ice and, of course, prices.[51]

The Knickerbocker Ice Company of Philadelphia, founded in 1841, eventually became the largest ice dealership in the country. As it grew over the course of the nineteenth century, Knickerbocker began to operate in multiple markets. It first made a name for itself in the interregional trade of the mid-nineteenth century, pioneering the same technology that would be used in the ice elevator for loading and unloading ships.[52] Knickerbocker expanded to the New York market shortly thereafter and controlled that market for twenty-five years. Managers made sure that its operations utilized all the current technologies in order to keep costs low. They also saved money through economies of scale. By 1880, it had become a vertically integrated operation with sites on the Hudson to cut the ice, barges to deliver it to docks in Manhattan, and a fleet of trucks to deliver it all around the city.[53] By 1888, nobody had more people cutting ice on the Hudson than Knickerbocker. The firm also had 700 wagons and teams of horses to distribute its ice throughout New York City and Brooklyn.[54] In 1891, it had sixty barges with an average capacity of 900 tons and sixty "double-ender steamers" to bring natural ice down from the parts of the Hudson north of the city where it was cut.[55] The power of the firm helped make New York City the largest market for natural ice in the country well beyond the turn of the twentieth century.

The Knickerbocker Ice Company formed the core of the American Ice Company, organized by Charles S. Morse, a shipping heir from Bath, Maine, in 1899. The "Ice Trust," as the New York papers called it, had a virtual monopoly on the market in both New York and Philadelphia. In the summer of 1900, the company doubled the price of ice for no obvious reason other than for profit's sake, a move which hurt the poor in particular.[56] That same year, the *New York Journal* revealed that New York's mayor and other members of his Tammany Hall political machine held many shares of the company. This created a huge scandal.[57] Despite New York State investigations and prosecutions, company founder Morse made $12 million dollars in profit when he sold his stock in 1900. Nobody in New York City ever went to jail for manipulating ice prices, but the scandal did damage a lot of reputations.[58] Constant headlines in local papers attacking the Ice Trust damaged the long-term viability of the Tammany Hall political machine. Mayor Robert Van Wyck was hounded by cries of "Ice! Ice! Ice!" during his public appearances. A reform ticket subsequently defeated him in his bid for reelection.[59]

Despite its impact on the politics of New York City, the Ice Trust spread throughout the country. In 1894, Knickerbocker erected its largest ice plant in the country in Philadelphia because it recognized that "under certain circumstances and contingencies the ice machine had become an actual necessity."[60] After its formation, the American Ice Company bought up all the ice plants in cities like Baltimore, Washington, D.C., and Philadelphia in order to gain control of those markets. Consumers in these cities used very little natural ice.

The proximity of New York City to the largest natural ice fields in the country helped that product survive longer there than elsewhere. The Ice Trust came to dominate the ice business there by buying up all the regional ice fields, especially along the Hudson River. Then they paid kickbacks to Tammany Hall so that no other ice company could unload their barges on municipal piers.[61] The kickbacks also kept mechanical ice out of New York City.[62] Even by 1915, with natural ice dealers around the country panicked by the competition of the machine-made product, the New York City market remained split approximately fifty-fifty between natural and artificial ice.[63] Since the trust determined the nature of the supply, people took whatever ice available. Lack of competition not only kept profits high; it also gave the Ice Trust little incentive to modernize its operations. Only changes in the national market eventually forced the New York City dealers to catch up with the rest of the country.

The Persistence of an Inferior Product

In response to the encroachment of machine-made ice, the Natural Ice Association of America issued a pamphlet around 1913 called *The Handwriting on the*

Wall. Subtitled "A Call to Arms!" the publication warned natural ice dealers that manufactured ice would soon destroy their business: "So long as the people of your community are in ignorance of the special advantages of natural ice, just so long will capitalists continue to invest money in ice making plants. For every ton of output of such plants you lose at least Two Dollars. For every ton of capacity of refrigerating equipment installed in your neighborhood you lose **Two Dollars More—PER DAY.**" The designers of the pamphlet wanted to enlist new members to join a publicity campaign in support of their product by rebutting charges that natural ice was unhealthy or inferior to manufactured ice in any way. "The time has arrived for definite action on your part," the association concluded. "Our work will be to educate the public. . . . You owe it to yourself to join forces with us all to preserve our markets in the years to come."[64]

Education, however, could only do so much. As the technology behind mechanical refrigeration improved, natural ice increasingly became unable to compete with artificial ice in terms of efficiency; and as the cold chain grew to become an increasingly important part of the refrigeration industry, efficiency mattered more and more. Although natural ice remained significantly cheaper than machine made ice, that advantage grew less important after the purity of the natural product got called into question. After 1910, machine-made ice began to rapidly force the older product from the market. Natural ice did have practical uses. Natural ice was still used in some iceboxes and for preserving food transported in railway cars, two essential links in the cold chain. Its low price made it a good product for preserving anything on a large scale. Because of such distinct advantages and disadvantages, natural ice and artificial ice could coexist at different points along the cold chain, but not indefinitely.

Predicting the imminent demise of the natural industry or predicting a bright future for mechanical refrigeration became a common refrain at the end of the nineteenth century. In 1894, for example, the superintendent of New York City's largest ice plant told the *New York Tribune*: "There is no doubt in my mind that artificial ice will reign supreme in a short time for the use of hospitals, hotels, families and all consumers who must have a pure and wholesome refrigerant."[65] Purity alone explains much of the popularity of artificial ice, but as the efficiency and reliability of ice machines improved, this feature became the primary explanation for why natural ice would inevitably disappear entirely. "It is safe to predict that the use of artificial ice for domestic purposes will entirely supercede the domestic product in a few years," argued the Triumph Ice Machine Company in its 1901–1902 catalog, "on account of its purity as

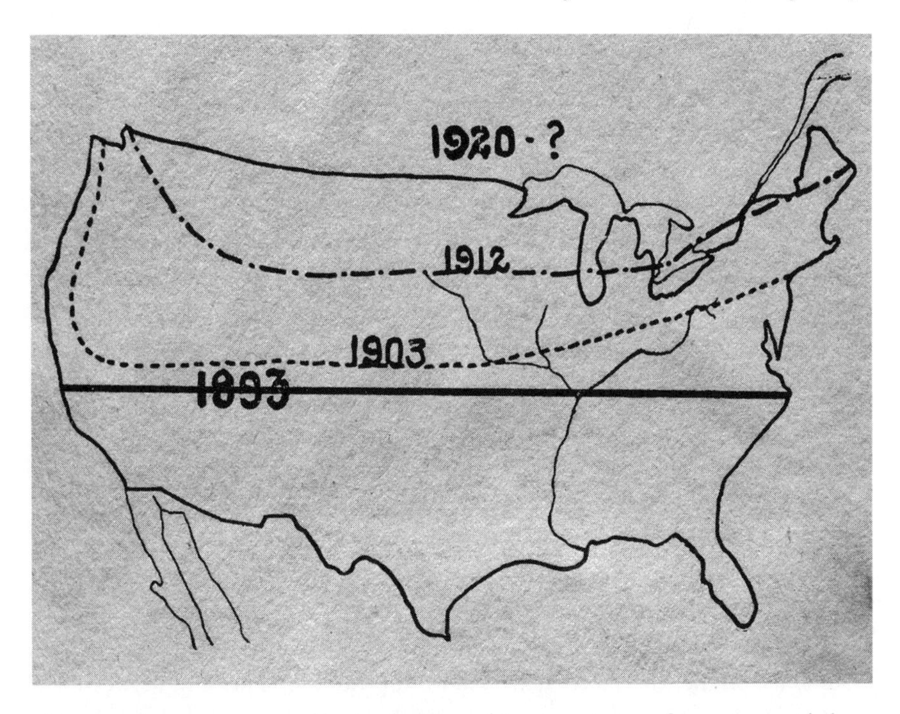

This map, from a 1913 pamphlet by the Natural Ice Association of America entitled *The Writing Is on the Wall*, illustrates the past and possible future spread of mechanical ice. The more efficient the machinery became, the further north it could successfully compete against natural ice. Lake Mohonk Mountain House Collection, box B-13. Courtesy of the Hagley Museum and Library.

well as its cheapness of production."[66] Sensing the increasing importance of the consumer ice market, a few large natural ice firms like Knickerbocker began to invest in refrigerating equipment for the first time during the mid-1890s but did not make the full transition until mechanical refrigeration technology improved and the advantages became increasingly clear. With these improvements, the cost advantage of natural ice gradually diminished. Yet despite the price drop for machine-made ice, the total volume of manufactured ice produced in America did not match that of natural ice harvested until 1914.[67] The inability of artificial ice manufacturers to ramp up production of machines to meet demand may explain the persistence of natural ice as much as the price difference between the two products. After all, nobody can switch to a product not available in their local market.

Natural ice could survive in direct competition with ice machines wherever the efficiency of the cold chain that supplied it allowed dealers to sell it at a lower price. Those conditions depended upon the interaction of technology and geography. Because of the huge volume of ice cut from the Hudson River, ice plants remained rare anywhere north of New York City at the turn of the twentieth century. In other words, the natural ice industry became a strictly regional business. "People from the West, as a rule, show no surprise when told that ice is made by machinery no matter how hot the weather," explained a Florida correspondent for the journal *Ice* in 1909. "The ignorant ones come from Massachusetts, Connecticut, Vermont and New Hampshire, the New England states which proudly think the world depends upon the natural ice cut from their lakes and streams."[68] It makes sense that the distance between the point of production and the point of consumption had shrunk over time because this helped keep costs lower. At the same time, the paucity of ice plants in the North kept the supply of machine-made ice low and hence more expensive.

In other places around the country, geography gave the price advantage to the artificial product. The example of New Orleans during the 1870s illustrates these factors well. Before the first mechanical ice plants arrived in that city, dealers sold ice on the wharf after it arrived from Boston for a price of between $20 and $30 per ton. Thirty to forty percent of that ice melted before it reached its intended customers. Though not particularly solid or hard, the first ice manufactured in New Orleans retailed for just $15 per ton.[69] When the city's ice consumers flocked to the cheaper product, the natural ice trade soon disappeared completely in that city. Over time, machine-made ice in markets across America became inexpensive enough that regular people everywhere could afford it. By 1900, natural ice had disappeared throughout the entire South. Elsewhere in the country, it persisted well into the 1920s, surprising most industry observers.

With natural ice unable to compete with artificial in terms of appearance, starting in the 1890s natural-ice manufacturers began to tout the point of origin for their ice as a way of signaling the purity of their product to consumers. In a way, this resembled the grading system used by the corn or cattle dealers. The industry wanted to train consumers to look for its label as a sign of quality and an assurance that its product would never be contaminated. No graduated scale of quality existed as it did for other commodities, but the natural ice producers wanted consumers to believe that all ice was not the same since some ice came from better water. "Pure Spring Water Ice from Ballston Lake," began an ad from the Ballston Spa Ice Company of Ballston Spa, New York: "Retails at 10 cents per 100 lbs. above any other ice," presumably because of its superior quality; "no sewage."[70] Abraham Rich printed his location in Gardiner, Maine, near the top of his banner ad in the August 1896 issue of *Ice Trade Journal*. The ad touted "heavy thick ICE (cut of 1895), of superior quality," alluding to the fact that older ice could keep more than one year in Maine (though its quality suffered over time).[71] A Chicago firm that cut ice from Lake Geneva in Wisconsin went so far as to publish the results of a chemical analysis of the water in its ice as part of its marketing efforts.[72]

This kind of point-of-origin advertising became increasingly important as the news about the safety of natural ice grew increasingly worse. "There has been much printed in the newspapers recently about the foulness of the Passaic River, and the many cases of typhoid which have resulted from the use of the water in Newark," reported the journal *Cold Storage* in 1899. "The water is so foul that the [ice-cutting] rights of the Simmons family are worthless, they say."[73] Yet judging from a complaint about the low price of ice in Newark in the same article, the Simmons family probably cut it anyway. This kind of behavior threatened the entire consumer sector of the natural ice industry and explains why, despite earlier industry claims that water purified itself through freezing, one natural ice dealer from New Jersey, in a 1909 speech entitled "What the Dealer in Natural Ice Must Do to Retain His Position in the Industry," recommended that dealers turn any firm that cut ice from polluted sources in to the authorities.[74] While the industry had once derided efforts by local governments to regulate its product, dealers of quality natural ice now welcomed these same efforts, in the same way that respectable meatpackers supported the Pure Food and Drug Act. A government seal of approval offered the prospect that consumers could buy their product without fear of adverse health effects. By 1910, industry representatives hatched a plan to certify the natural ice supply themselves, but by then the poor reputation of their product had gotten stuck in the public mind.[75]

While their natural-ice competitors struggled, machine-made ice companies increasingly used health concerns directed at *all* natural ice as a marketing tool. "Your physician will tell you that a great deal of sickness is caused by drinking impure water," wrote one ice company in an ad from 1909. "Impure ice contaminates not only water alone, but all the food with which it comes in contact."[76] By this time, iceboxes had become an important household use for ice. The possibility of dirty water touching the food in these containers—literally boxes with ice in them—posed a significant threat to people's health. The larger the consumer market a firm had, the more likely it exclusively handled manufactured ice. "The character of the lake ice which has been found in my own ice chest has been bad," explained one of seven physicians whose testimonials appeared in a 1914 advertisement for artificial ice, "and a menace to the health of the community."[77] Point-of-origin marketing essentially said, "You may have read about health scares related to pollution in some waters, but that does not apply to the pristine body where we cut our ice." That argument could only go so far because of the physical appearance of so much natural ice. Despite this disadvantage, even dirty natural ice, chock full of sediment, successfully competed against artificial ice for over twenty years on the basis of price. People who could not afford the best quality ice still bought natural ice.

Technological change came slowly to the American ice industry. An entire country of ice consumers that included rich and poor alike meant that consumers in different segments of the market would choose their type of ice for different reasons. People who could afford it generally chose artificial ice when it was available. Consumers who had to stretch in order to afford ice at all generally chose natural ice because of its price. Industrial users generally chose natural ice for the same reason. (Consumers in other countries like Great Britain and France did not have to make the choice between natural and artificial ice because refrigerating engineers had already perfected mechanical refrigeration by the time they developed a taste for this product.) As the technology evolved and artificial refrigeration became more efficient and reliable, artificial ice became cheaper. Only then did natural ice sprint towards extinction. New technology did not determine the fate of old technologies in this instance. The reaction of consumers did.

More efficient methods of harvesting and housing ice developed after 1890 helped make natural ice more competitive for a few years; but after 1910, despite a brief comeback during World War I due to a shortage of ammonia, the natural ice industry declined quickly.[78] In retrospect, it seems inevitable that mechanical refrigeration would lead to the natural ice industry's effective demise. In 1921, only one icehouse remained on the Kennebec in Maine.[79] By 1928, the natural ice industry had almost faded into history. It constituted only a small portion of the total ice sold, and its consumption occurred mostly in parts of the country with long winters.[80] By the 1930s, the spread of the household refrigerator, in which consumers could make their own ice, severely curtailed the market for both the natural and the machine-made product in the way it had always been supplied.

The economic model that each segment of the industry depended upon to grow ensured that artificial ice would win out in the end. The natural ice industry, as Thoreau had joked, really did have more in common with farming than it did with manufacturing. Entrepreneurs like Tudor could do little things to improve the quality of the ice they harvested, but the size and quality of their "crop" was at the mercy of outside forces like the weather or the number of environmental polluters in the vicinity. More importantly, while mechanization allowed them to harvest ice faster, the only way they could scale up production and get more ice than nature created was to find more bodies of water from which to extract it. The artificial ice industry, on the other hand, followed the economic model common to any mass production industry. As its technology improved, it could create more ice and provide it to a growing market on an increasingly regular basis. While both segments of this industry faced transportation-related bottlenecks, whatever advantages natural ice once had must have disappeared as the price difference between the two products closed.

The growing market for ice of all kinds in the growing cold chain allowed both segments of the industry to flourish for a time. The uses of ice and refrigeration in general expanded as the technology for making/harvesting and transporting both products improved. Most of these new uses involved preserving food, either in transit from place to place or in the same place over an extended period of time. Since both natural and artificial ice melted, they both required replenishment. Over time, the general adoption of refrigeration for perishable food meant that ice was replaced by mechanical refrigeration all along the many supply chains that brought groceries of all kinds to market. By making it possible to ship perishable food from all parts of the country and the globe or to store seasonal food throughout the year, first ice and then refrigeration changed what the world eats by allowing people to defy nature. To do this required moving perishable food without having it spoil. Making that possible took the development of more new technologies over a period of decades.

CHAPTER 4

Refrigerated Transport Near and Far

In 1898 the commander of the U.S. Army, General Nelson A. Miles, charged that Chicago packinghouses had sold the government condemned meat to feed the troops during the Spanish-American War. Some of the questionable meat came in cans. Some of it came in a refrigerated state. Miles claimed that packers had preserved 337 tons of meat from this second category by "embalming" it. That meant the packers had allegedly injected it with dangerous chemicals. In response to these charges, President McKinley appointed a commission headed by Major-General James F. Wade to investigate. That investigation found no evidence of any chemically tainted beef. Instead, Wade's commission blamed the way that the packers and shippers handled the meat between Chicago and the soldiers in the field. Poor sanitation, slow delivery, lack of ice for preservation en route, lack of cold storage warehouses in Cuba—these holes in the cold chain caused the problem which had led to the complaints.[1] "I ate of the beef which the transport *Manitoba* brought to Porto Rico [*sic*]," explained Captain Barclay F. Warburton in testimony to the commission. "It was so bad that it was impossible to swallow it. . . . I had no idea that this meat had been subjected to any chemical process, but believed the beef to have decomposed on account of the lack of proper refrigerating facilities caused by the clogging of the machinery used for that purpose on the *Manitoba*."[2] While this wartime debacle often gets categorized as part of the fight for a national pure food law, it also serves as an excellent illustration of the importance of an effective cold chain.

A successful cold chain has to set perishable food in motion and keep it fresh over long periods of time in transit. This chapter explains the costs and benefits of transporting foods over long distances. The next one considers cold storage, which keeps food cold in place over extended periods of time. All told, wrote Artemus Ward in *The Grocer's Encyclopedia* from 1911, these two developments "rendered possible uniform distribution of fresh foods throughout every part of the country and carried the surplus, not only of America but also of Australia and Russia, to the markets of Europe."[3] Refrigerated transport allowed the surpluses created by industrialized agriculture to sell globally for the first time. This led to decreased prices for perishable goods, finally making many such products affordable to ordinary consumers. Other perishable products (such as bananas), sold in places where they could not grow normally, defied the seasons in a different way; they could be sold where they had not appeared before. As a result, even countries with few iceboxes or mechanical refrigerators still benefited from long cold chains. People could still go down to their local markets daily to buy perishables from distant lands even if they had nothing at home to prevent the food from spoiling once it arrived.

Perishable food became an article of commerce long before the ice and refrigeration industries perfected the cold chain. Sending pickled or canned food along trade routes required few special considerations, but these kinds of foods did not have to stay fresh. As an article of commerce, ice traveled from one end of the earth to another without refrigeration even though much of it inevitably melted. But perishable food preserved by ice could travel only short distances without significant deterioration. It often spoiled because other factors besides low temperature—such as the humidity of its storage place or whether it stayed at the right temperature—determined whether it would stay fresh. That explains why maintaining precise temperatures along the cold chain became so important. Because the perfection of mechanical refrigeration allowed for this kind of control, it greatly improved the success rate for refrigerated transport. That greatly spurred the development of new forms of mechanical refrigeration that eventually became both effective and cost efficient.

At first, eating food grown far away seemed like a miracle. In the past, people did not generally see perishable food transported by ship or by railcar, but they still understood that their out-of-season fruit originated elsewhere. As the technology of refrigerated transport made longer cold chains possible, the distance that most perishable foods traveled in America grew. Writing in 1900, Theodore Dreiser noted that all the strawberries consumed in the greater New York area had been grown on Long Island and in New Jersey only twenty years before. Then the range expanded to Delaware, Maryland, and Virginia. By 1900, strawberries were grown in the South throughout the year and sent to northeastern markets by specially routed fast freight trains. "One must see the depots and the trains, the flourishing gardens, blooming in January in Southern Florida," Dreiser wrote. Despite the distance traveled, it cost only two cents to ship a box of strawberries from Arkansas to New York City. The low shipping cost meant it was much easier to import foods than to build a multitude of greenhouses in which to grow New York's fruits and vegetables.[4] Did shipped strawberries taste as good as the strawberries grown close by in season? Probably not. Nonetheless, lengthening cold chains made it possible for other regions of the country to make best use of advantages like a warm climate or surplus open space, while simultaneously allowing the price of many such goods to drop. Despite increased efficiencies, strawberries in winter retained the ability to surprise people for many years after 1900, and people happily paid for that privilege.

As time passed and the separation between the point of production and the point of consumption grew, many consumers lost whatever understanding they once had about where food came from or the way farmers produced it. Only when their meat got spoiled en route did people worry about how the cold chain worked. On the other hand, when refrigerated transport worked well, it brought what seemed like an infinite variety to a very monotonous American diet. Instead of just canning or freezing foods grown in your area, you could buy nearly anything imported from nearly anywhere. As one food analyst wrote in 1950,

> From the food stores the 1900 housewife could buy such staples as sugar, rice, beans and a few canned fruits and vegetables. She probably was able to secure dried fruits. She may have been able to buy meats, or she may have had to settle for mackerel. If she had any fresh vegetables, she may have had her choice of such items as carrots, turnips, and cabbage, which she had stored in her cellar. The same housewife in 1949 had her choice of every conceivable type of fresh fruit or vegetable, or canned fruit or vegetable. The 1900 housewife was almost entirely dependent on products that had been grown in her immediate vicinity, whereas the 1949 housewife had her choice of products grown anywhere in the world.[5]

The development of the modern cold chain made this amazing transformation possible. Before the advent of the household refrigerator, cold itself traveled from place to place in the form of ice. Because ice melts, its distribution created more complications than happened in the distribution of just about any other commodity. Icemen found an important place in American economic and cultural life during the early years of the ice industry because they serve as the most visible symbol of that industry.

"Thank God for Refrigerators"

Eugene O'Neill's stage classic *The Iceman Cometh* may be the only reason most people still remember ice delivery men. While the archaic language of the play's title evokes the Bible, it also suggests a dirty joke still popular in 1939 when O'Neill wrote it. The humor in that joke depended upon the common stereotype of an iceman's strength and virility.[6] Bernard Liquie, an ice deliveryman in Detroit during the 1930s, described his job on National Public Radio in 2010: "Twenty-five pounds you put in your hooks. Fifty pounds you put it on your shoulder. You wore a wool shirt so that it don't go through to your shoulder blade and freeze you to death. Yeah, we took ice upstairs and downstairs, into basements—wherever it took. Made you tired. . . . Damn hard work. Thank God for refrigerators."[7] The boxer Joe Louis built up his muscles as a young man while delivering ice in the same town. When football Hall-of-Famer Red Grange was in college, fans gave him the nickname the "Wheaton Iceman" for doing the same thing while growing up in that Illinois town.[8] Ice companies had to pay dearly for the services of men strong enough to perform this job. A former iceman in Chicago remembered that he received $18.00 per week in 1918 and $30.00 per week the next summer.[9] A 1922 "treatise" on selling ice estimated that 26 percent of the cost of ice at retail went to the wages of the deliverymen. The next biggest components of the price were the expenses for stabling the horses (22.1%) and the cost of making the ice itself (22%).[10]

The first icemen appeared on the streets of American cities in the early 1830s. By the 1840s, they had become a fixture of urban life.[11] Whether a firm cut ice from a lake or manufactured it with machinery, someone still had to deliver it because the vast majority of customers could not pick up something so heavy and bulky. Many small firms acted in a double capacity as ice and coal delivery services, since coal demand picked up just about the same time of year that demand for ice slackened. The ethics of ice deliverymen was a persistent problem in the industry. As one ice plant owner explained in 1903, "I take them just as they come and make the best of them. If they don't suit, I fire them. If I catch them stealing, I fire them."[12] Ice companies constantly worried whether their icemen shortchanged their customers on the weight of their ice and pocketed the proceeds. Other drivers sold a little ice on the side, chalking the difference up to melting. Distrust for their own drivers explains why many firms urged their customers to get their ice weighed upon delivery.[13] Having customers buy

John Burbank, iceman. In the heyday of the American ice industry, the iceman with his cart of either natural or artificial ice was an iconic image in cities across the United States. Courtesy of Denver Public Library, Western History Collection, Photo X-10259.

coupons in advance of ice purchases not only made it easier for ice distributors to know how much to buy from a manufacturer; it also meant that deliverymen were not dealing with cash. Sometimes ice companies instructed their drivers to visit customers to sell coupon books when not delivering ice. Most offered their drivers a commission for each book sold.[14]

Dealers in natural and artificial ice competed fiercely in offering polite service because the quality of delivery often determined from whom customers bought their ice. "Through you and YOUR ACTIONS many gain their impressions and form their opinion of the Company for which you work," suggested the Utica, New York, Ice Company to its deliverymen. "Indifference, loss of temper, manner and language unbecoming a gentleman WILL NOT BE TOLERATED."[15] Convenience and efficient delivery served as another selling point for attracting business. "Our ICE TRUCKS deliver on schedule[d] time daily, except on Sundays," explained an ice card for the J. F. Bell & Sons Company of Berlin, New Hampshire, from the early years of the twentieth century. Generally, customers put a card furnished by the company in a window to indicate that they wanted a delivery that day. Delivery firms signed contracts with customers that specified a predetermined amount of ice each week for a particular season, but J. F. Bell made it easier to make sure that each customer got only as much ice as he needed and wanted: depending on which side of the sign was at the top in the window, the customer could indicate whether he wanted a fifteen-cent, twenty-cent, twenty-five-cent, or a thirty-cent piece of ice.[16]

The lack of such delivery services retarded the growth of ice consumption in other countries around the world, even in countries that had large ice industries geared towards industrial users. In Great Britain, the Wenham Lake Ice Company delivered ice to houses around London and sent it by ordinary freight until at least 1874, but some time after that, delivery service disappeared.[17] At the turn of the century, Englishmen had to go to fish stores to find ice they could use at home. "It is absurd," opined the medical journal *The Lancet*, "that practically the only shop where [ice] may be obtained is that kept by the fishmonger, who keeps it not for purposes of consumption but for cooling fish. . . . It seems ridiculous that in the English summer the demand for ice should not be sufficient to warrant the establishment of a special agency for its supply."[18] The *Lancet*'s editorial staff did not even know how Americans received their ice, a clear indication that the delivery services that had once existed in England had disappeared by this time.[19] In other countries, the delivery systems fell even further below American standards. In Venice, for example, ice deliverymen carried their product by gondola with a box on their back that held only eight pounds. "It is not so easy," said *Ice and Refrigeration*, "to imagine how these peripatetic ice men would supply a fete day or even a moderate sized modern American bar."[20] Since Americans consumed so much more ice than Italians did, they required more efficient cold chains.

As ice itself became a smaller and smaller part of the cold chain, ice delivery became an increasingly less significant part of the ice business. In the classic Marx Brothers movie *Horse Feathers* (1932), Chico and Harpo Marx play ice deliverymen. As the movie opens, Chico is doubling as a bootlegger, while Harpo's character also serves as the local dog catcher. Their characters needed multiple professions because ice delivery had collapsed by the late 1920s. That explains why the largest ice plant in the world belonged not to an ice manufacturer but to a railroad, the Pacific Fruit Express.[21]

Mechanical refrigeration worked much better in large seagoing vessels than smaller railroad cars because the cars could not hold bulky mechanical refrigeration equipment. The need to transport many kinds of foods over varying distances meant that shippers had to address and resolve a variety of problems specific to both of these situations. Many global cold chains benefited countries (like Australia) that had virtually no internal cold chain at all.

Across the Ocean

At the beginning of the nineteenth century, ice was merely a commodity. The feat of shipping New England ice to tropical climes, while impressive on its face, made little money for the people who accomplished it because only the affluent could afford to buy ice at the necessary price after such costly voyages. This explains why ships carrying ice always carried at least some food packed in with their primary cargo.[22] Since the preservation of food depended more upon environmental conditions than the ice itself did, shipping something besides ice never worked well when ice served as the only source of cold on the journey. Travel through a warm stretch on a journey, and a boat packed with ice simply lost more ice than usual. If the fruit or vegetables packed in that ice faced such conditions, the crew had no way to save them. As a result, large-scale refrigerated transport of perishable food only came later in the history of the cold chain, after mechanical refrigeration became viable.

During the 1850s, large luxury ocean liners—"floating palaces," people called them—kept fresh fish, food, and produce on ice while crossing the Atlantic.[23] This food, however, was solely for the benefit of the rich passengers during the weeks-long transatlantic crossing, rather than for the people at the ship's port of destination. Not until the 1870s could firms ship enough perishable food reliably enough to keep the price down so that more than just the wealthy could afford to buy it. Mechanical refrigeration played a huge part in this change. Not only did it keep food better than ice, but it also did not take up nearly as much volume as ice did on the journey, leaving more room for more food. The development of the transoceanic trade in perishable food proved particularly important to highly isolated countries because they could now sell many of their agricultural products in the world market on a mass scale, thereby facilitating substantial economic development and the migration of European peoples needed to keep that development going. This proved particularly true for Australia.

In July 1875, the Australian industrialist Thomas Sutcliffe Mort addressed the Agricultural Society of New South Wales about his plans to transport frozen meat from Australia to London.[24] After laying out the technological challenges involved, Mort explained his assessment of the supply and demand for meat in Australia and Great Britain. He noted the abundance of livestock in four Australian states (New South Wales, Victoria, South Australia, and Queensland) relative to the number of people there: in 1874, there were 1.7 million people in these four colonies, compared with 5.6 million cattle and 47.8 million sheep. According to Mort's calculations, which allowed for each person to consume 350 pounds of meat per year, there was an excess of nearly 300,000 tons of meat in Australia. Britain, on the other hand, Mort explained, faced a terrible shortage of fresh meat for the masses. Many people there had expressed a great concern as to whether the meat supply there would be able to keep up with the growing population.[25]

As the *Times* of London explained the problem (in an article that Mort cited in his speech), "There are parts of the world adapted for maintaining vast herds of cattle and sheep, and adapted for no other purpose, and from these the more densely populated countries might be permanently supplied with sustenance"—assuming, of course, that somebody could develop a means to get fresh meat across the world without spoiling.[26] Mort believed he could do that. And he did not stop with meat. Mort proposed that Australia's oysters—right there for the taking in Sydney's harbors—could be shipped to England at an immense profit. He had already begun to freeze fish for local consumption and proposed exporting them too. Mort also outlined how to chill fresh milk from the countryside with mechanical refrigeration for urban consumption.[27] His conception of such a lengthy cold chain served as an important precursor to the multistep, global cold chains of today.

On 8 July 1877, after years of experiments, Thomas Mort's firm packed its first trial shipment of lamb onto a converted iron barge, the *Northam*, to take it from Sydney to London. His decision to begin with a large initial shipment meant committing enormous financial resources for an experiment. Even for a man of Mort's wealth, this was a gamble because mechanical refrigeration was not yet perfectly reliable. Unfortunately for Mort, this experiment failed before it ever got under way. The report of the board of directors of Mort's firm, the Sydney Fresh Food and Ice Company, to stockholders explained the situation this way: "On the 9th, the refrigerating apparatus was to all appearances working satisfactorily, having reduced the meat chamber temperature from 52° to

35° [Fahrenheit] in 60 hours, but in the afternoon of the same day an accident occurred through the action of the nitrate of ammonia on the coils . . . which caused immediate stoppage of operation."[28] The firm removed the meat from the ship before it sailed. The failure of Mort's experimental shipment harmed his health and accelerated his death from pneumonia in 1878.[29]

Many people besides Mort tried to harness the power of refrigeration for this kind of commerce. Between 1849 and 1869, various countries around the world issued 2,000 patents related to the preservation of meat, since ice damaged meat even when not in direct contact with it.[30] In 1868, Argentina offered a 40,000-franc prize to whomever could invent the best way of preserving fresh meat. Nobody ever collected it.[31] Mechanical refrigeration on board a ship presented many problems not faced on land. For example, shipboard refrigeration had to be reliable throughout months-long ocean voyages when replacement parts would be unavailable. Engineers also had to create machinery that could withstand salty sea air. Even if refrigerating units took up less room than ice, the same space-saving principle applied here too. The bigger the machine, the less space left on board for meat.

Since Texas, like Australia, had a huge surplus of beef, the Glasgow firm of John Bell and Sons began experimenting with shipments of live cattle to Great Britain during the 1860s. Because these animals grew scrawny on the trip, Bell switched to dressed meat (meaning that cattle had already been slaughtered and prepared for purchase at the slaughterhouse) for the first time in 1875, using a process invented by T. C. Eastman of New York that used ice as a refrigerant. All of the meat came through the journey without spoiling thanks to this new process.[32] The Frenchman Charles Tellier pioneered the successful use of mechanical refrigeration to preserve meat during transocean voyages in 1876, but that cargo was severely damaged during the trip.[33] In 1878, the first successful shipment of frozen meat in the history of the world went from South America to France.[34] By 1885, the shipments of meat from the United States to Great Britain traffic had increased to 30,000 tons per year, aided by the switch from ice to mechanical refrigeration.[35]

American beef found a large market in Great Britain even before the transformation from ice to mechanical refrigeration because it could undersell the British product in Britain thanks to the resources and infrastructure at the control of the Chicago packinghouses. The *Standard* of London reported in February 1877 that it had been "fifteen months since [American] consignments reached the central meat market of Smithfield, and they have continued in continually increasing proportions ever since. All this meat has fetched high prices, the greater part being purchased by West-end butchers, whose customers have consumed it in happy ignorance of its origin."[36] With such results, many English cattlemen panicked. "Even when beef is at 10d. per pound there is not much profit in it," wrote one farmer to the *York Herald* in 1877. "If we have to compete with beef at 6d. per pound we shall be beaten out of the market, as no farmer can afford to feed beef at that price. . . . If much of this American beef comes into our English markets I can foresee that England will no longer be a beef-producing country."[37] American beef captured 94 percent of the British market between 1880 and 1889, its most successful period.[38] Some of that figure represents live cattle shipped along the same route and slaughtered upon arrival, but assuming the cows even survived the trip, the loss of weight experienced on the trip made this method too expensive. Live cattle (unlike frozen meat) also had the potential for spreading disease. Because of these problems, the live cattle trade declined sharply as shipboard mechanical refrigeration steadily improved.[39]

American beef captured the English market so quickly because of its low price and high quality.[40] Both these traits derived from the technology that preserved it. American producers kept their beef chilled rather than frozen for the comparatively short voyage across the Atlantic. English observers claimed that butchers often substituted American beef for British beef in order to gain a price premium and avoid a stigma that some consumers had against buying anything from the United States.[41] Yet most domestic English beef seemed thin by comparison. Salted or canned beef, two other alternatives, simply could not hold a candle to fresh meat and quickly became the exclusive province of those who could not afford the better kinds.

The beef and mutton from Australasia or South America, which always arrived frozen until the years immediately preceding World War I, could never compete on quality because of the damage freezing did to the product. Freezing affects the physical structure of meat, making it less moist. It can also rupture small blood vessels in the muscle fiber, making it look less appetizing.[42] *Ice and Refrigeration* described the pitfalls of transoceanic frozen meat transportation this way in 1902: "The chief danger is in freezing the meat too rapidly, so that a coat or wall of frozen material (virtually insulating material) is formed around the warmer interior part of the meat, driving the heat further and further toward the center. Thus gases are imprisoned, decay of bone and marrow sets in, and when the meat is thawed the interior foulness becomes evident."[43] Frozen meat also created a problem because meltwater formed en route could leave the meat discolored. Such problems greatly hindered the development of the Australian meat export industry, which had to freeze its product because the trip to Britain was so long. Since so many shipments from Australia were unmarketable by the time they reached Britain, Australian exporters had to pay huge insurance premiums for each voyage. This drove up their costs, and this, in turn, eventually priced their product out of the market as they faced greater competition.[44] Distance also constituted a real barrier to delivering high-quality meat around the world. Various Australian firms tried five times to keep meat chilled rather than frozen on its way to England starting in 1895, but did not succeed until 1909. (Even then, shippers sprayed the meat with formaldehyde to preserve it during the long voyage.)[45]

Like meat, the transport of fruits from outside the United States into the country via refrigerated vessels also originated during this era. Bananas defied the seasons before any other fruit even though they had a longer distance to travel than most. Enterprising traders shipped bananas year round from the tropics to American port cities.[46] Bananas, like many other fruits, can survive without refrigeration. There was nevertheless substantial loss from spoilage. In 1885, a twenty-five-year-old produce buyer named Andrew Preston had a better idea: he began buying green bananas and preserving them on vessels with ice to retard ripening. To get his bananas from the tropics to the table, Preston created a network of refrigerated cargo vessels, boxcars, and storerooms around the country. With the help of the businessman and adventurer Minor Keith, the two formed the United Fruit Company in 1899. Another importer, Joseph Vaccaro of New Orleans, supposedly bought up every ice factory on the Gulf Coast to service his banana supply network. That company went on to become Dole.[47] Because of the success of firms like these, the price of bananas dropped from five or ten cents each to ten or fifteen cents a dozen from the late 1800s to the early 1900s.[48]

Refrigeration turned bananas from a luxury to an extraordinarily common food despite their distant points of origin. United Fruit introduced its first mechanically refrigerated vessels in 1903. Unlike meat, which traveled in refrigerated holds, bananas were cooled only by huge fans, which circulated fresh air to keep the cargo cold (or to circulate heat if the bananas arrived during winter).[49] By 1914, fifty million bunches entered American ports each year. At first, these bananas were available only in and around port cities, but by 1893 producers began to ship them inland. Thanks to "well-organized" rail service from New Orleans to Chicago, they sold for less in Chicago than they did in

New York City at that time.[50] The banana market could not have existed without the speedy rail service because bananas could not survive long journeys without some control over temperature. When stored in too warm a room, they ripen quickly and will not stop, causing shippers to dump them all as fast as possible at a price too low to recoup their costs.[51] When bananas are stored below 45 degrees Fahrenheit for any length of time, they go black and are then substantially harder to sell. This made it especially hard to ship bananas in winter unless the shipping compartments had oil stoves.[52]

Despite such difficulties, the example of the banana shows how refrigerated transport could change the diets of even the poorest people. By bringing new foods, particularly fruits, to regions of the world where they did not grow naturally, refrigerated transport enabled people to eat healthful foods that had previously been unavailable. Because these new or rare foods were shipped in large volumes, grocers could offer them at a lower price than ever before, sometimes even at a lower price than locally grown fruit in season.

The effectiveness of refrigerated ocean shipping has only improved in recent decades. A New Jersey trucking company owner named Malcolm McLean sent the first container ship on a journey from Newark to Houston in 1956. Unnoticed at the time, container technology has greatly improved the efficiency of the transportation of nearly everything. Since cranes can now lift these boxes from a ship and place them directly on a truck, the shipping container made it possible to bypass high-paid unionized stevedores. This and other container-related increases in efficiency have caused the cost of ocean shipping to drop precipitously as these containers have become widespread.[53]

The refrigerated shipping container was a natural result of the move from railroads to trucking. Once you could refrigerate a truck on the road, it became possible to refrigerate containers on the ocean, reinforcing the sides so that containers could be picked up by cranes and stacked on top of each other.[54] By 1973, these units had become fully automated.[55] In the late 1970s, Barbara Pratt, of the shipping firm Maersk, worked and lived in a shipping container in order to study issues like how air flowed through them. Her findings led to a complete redesign of refrigerated containers and even greater efficiency.[56] Today's refrigerated shipping containers have all the advantages of regular shipping containers except that they also have sophisticated refrigeration systems designed to preserve perishable food (or any other commodity that requires refrigeration). These mobile units are better than ever because their complex, computer-controlled systems can regulate not only temperature, but also humidity, ventilation, and gas levels. Since these units are self-contained, they can be mixed in with nonrefrigerated containers on ships carrying a huge variety of all kinds of cargo. The newest models can even send status reports about the cargo wherever it is in the world via satellite.[57] As of 2000, about 550,000 marine containers transported food throughout the world to approximately 1.2 million refrigerated road vehicles.[58] This technology more than any other has made it possible for cold chains that transport a wide variety of products to become truly global in reach.

Across the Country

The same cold chain that allowed for shipments of perishable food across the oceans also allowed large Chicago meatpackers to underprice local butchers on the east coast of the United States with meat slaughtered a great distance away. Anyone who wanted to ship anything by rail in late-nineteenth-century America faced many difficulties. Going from point A to point B might require several stops accompanied by long waits because the various sections of long routes were likely to be owned by different railroad lines. Prices along these routes could vary widely, with local routes sometimes costing substantially more than long-distance routes. Scheduling the arrival of cargo was also a problem. Most freight trains were unscheduled, ideally leaving only when most of the cars were filled to capacity. Such scheduling delays, or bad weather, were obviously an issue for perishable food shipments, since delays could lead to the cargoes of entire train cars becoming unfit to eat. Once the shipment of perishable food out of California became possible, it had to go on special fast freight lines, costing more than nonperishable cargo of a similar weight and volume.[59] Because of these and other transportation-related difficulties, it took until the 1880s and 1890s to develop a viable railroad car to ship large volumes of fruit out of that state.

Most other countries around the world did not have even a poorly operated transportation infrastructure for perishable food. In Great Britain during the late 1870s, imported meat had to be dumped on the London market, sometimes at a loss if the weather was hot, because there were not enough refrigerated railcars to distribute meat from London throughout the rest of the country.[60] As late as 1891, the American consul in Sydney detailed the ordeal of live cattle from the Australian interior to slaughter near the docks. They spent over a week without any food and sometimes without water, so that "the cattle may be seen at the abattoirs with their heads hanging down, their bellies tucked up to their backs and looking utterly miserable and wretched." As a result, "it is notorious [that] the meat supplied in Sydney is, as a rule, inferior, tough, and void of flavor, especially in the bad seasons." To combat this problem, Australia had just begun to experiment with refrigerated railway cars at the time of the consul's report.[61]

By that time, the United States had been experimenting with the same technology for approximately thirty years. Before the late nineteenth century, most cattle were raised in the vicinity of slaughtering houses because live cattle obviously did not decay. Local butchers around the country killed them in small-scale operations and sold their meat quickly before it spoiled. Brining or canning beef offered alternatives, but since the taste was so unsatisfactory, these did not impact the market for fresh beef.[62] The first experiments with bringing western meat to eastern markets involved the shipment of live cattle. This was not only inefficient, since it meant that shippers had to pay to transport parts of the cattle that were unfit for human consumption, but it also affected the tastier parts of the cow too. Live cattle transported from Cheyenne to Chicago lost, on average, 100 pounds because they were usually too scared to eat during the trip.[63] These cattle also required expensive feed bins and stockyards for rest over the course of such long journeys.[64] Otherwise, they would have lost even more weight.

The transportation of dead (and therefore perishable) food created even bigger problems for railroad companies than it did for the ice and refrigeration industries. Once the intercontinental trade in ice ebbed, independent entrepreneurs and railroads handled the movement in the cold chain across space. As mechanical refrigeration spread, entrepreneurs used that chain more often for food than for ice. Ice then became the source of refrigeration that kept that food cool rather than a commodity itself. Railroads that operated refrigerator cars became customers for refrigerating machinery to use in storage warehouses. Ice cost less than mechanical refrigeration for the purpose of shipping food, but it still required a lot of infrastructure too. Ice became the only practical way to get meat and other perishables across the country after packers built a series of icing stations along the way to refill refrigerators cars when too much of the initial ice placed in the car had melted. In this manner, producers could concentrate their slaughter operations in Chicago. They could also break their cows into pieces and ship the parts with pinpoint accuracy only to those places that wanted them, thereby cutting down on waste. One steer or pig became multiple commodities, all of which the packers shipped separately from the rest of the animal via refrigerated railcars.[65]

These kinds of efficiencies explain why many people tried to invent the refrigerated railcar at the same time. They all recognized the obvious value of a successful model.[66] Despite experiments in the early days of American railroads, the real growth in the use of this technology came after the Civil War, when packers made various attempts to ship meat across the country. The first American patent for a refrigerated railcar was issued in 1867, but many prototypes existed around this time.[67] There was a Tiffany patent, a Zimmerman patent, an Anderson car, and a Wickes car; but the basic design of all these cars were similar, since any car could work to some degree if you packed enough ice in it.[68] The goal was to create a car that used as little ice as possible while still keeping as much meat as possible fresh. This strategy kept costs down and profits up but did not prove entirely satisfactory. Meat could still get damaged even if it did not come into direct contact with ice. Just moist air could blacken 8–10 percent of the surface of the meat.[69]

The packer Gustavus Swift made the most successful and best remembered of these efforts towards perfecting the refrigerated railcar. He hired an engineer named Andrew Chase to work from a patent taken out by Arnold Zimmerman that emphasized the importance of ventilation inside the car. Chase included some of his own innovations and created a very effective car.[70] Building a fleet of refrigerator cars cost lots of money, so railroads declined to do so initially—not just because of their expense but because dressed beef undercut the live cattle market which railroads found so lucrative. Once the fleet was built, if Swift could not ship enough beef east to meet the demand his low prices and advertising created, all the innovations would be for nothing. By 1883, Swift had set up enough icing stations along his northern rail route to resupply natural ice in his cars and keep his perishable cargoes cold all the way to the end of their long journey.[71] This too proved expensive. Despite high costs, the mass production of dressed beef in Chicago beginning in the 1860s justified the investment in a network of icing stations to preserve the cargo on its journey east and still made it possible for Swift to undersell eastern slaughterhouses while profiting from each sale.[72]

As his firm grew larger, Swift both shortened and reinforced the effectiveness of his cold chain. Swift's son later wrote about his father: "It was toward the close of the '80s that he raised the question of building branch plants still nearer the source of supply than Chicago. Beef cattle were coming principally from the West and Southwest. Why not slaughter them near their points of origin and thus effect savings comparable to the savings which had been attained when beef was dressed at Chicago instead of at Fall River?"[73] Large packers like Swift developed effective distribution centers with refrigerated branch houses and "peddler cars" that distributed small quantities of meat to smaller cities.[74] Swift even delivered his meat in iced wagons.[75] He used mechanical refrigeration only in abattoirs and for on-site storage, since those machines were still too big for any link of the cold chain except ships. This did not hurt Swift's bottom line because ice (first natural, then manufactured) cost so little. As a result, no financial incentive existed to develop mechanical refrigeration for railways until many years later.

Even with this simple technology, successfully shipping meat across the country still required a lot of capital. The expense of the infrastructure to operate refrigerated cars, most notably large icing facilities, helped to stimulate the growth of the meatpacking industry into an oligopoly. The efficiencies needed in order to operate these cars effectively also explains why packers concentrated their slaughtering in Chicago. If more cattle were slaughtered faster, no railcar had to leave town with any empty space in it. Those efficiencies helped pay for the high initial costs. In 1883, for example, a single boxcar cost $800, while a refrigerator car—with its insulation, double walls, and other attributes designed to conserve ice—cost 50 percent more.[76] In 1910, it cost twice as much to build and maintain a refrigerator car as it did an ordinary boxcar, since refrigerator cars required refurbishment before every trip due to water damage from the melting ice.[77] (And if the railroad mixed salt with the ice in order to make the car's temperature drop faster, this caused even more damage to the car and the railroad tracks.) Such costs promoted the consolidation of the industry, as only the largest firms could make such investments. By 1918, the five largest packers in the United States owned 90 percent of the railroad routes connecting packing houses to the cold storage stations that served as a vital stop in the distribution of their product. Independent packers were almost entirely shut out of this business.[78]

Seafood traveled by rail for the first time in 1842, years before anyone imagined the dressed beef trade. The first such cargo was a single live New England lobster. Lobster meat spoils quickly after the animal is killed, so when this lobster died en route, it got cooked in Cleveland and shipped the rest of the way on ice.[79] Before any seafood made it across the country, however, it first had to make it back to port. Unable to carry their catch long distances in a fresh-caught state, fishermen used ice to completely freeze fish once they reached shore. The first fishing boat to carry ice on it for preservation departed Gloucester in 1838. Initially, fisherman separated the ice from the fish, but that changed when they discovered that ice, as it melts, can keep the skin of the fish moist.[80] Keeping a fish on ice preserves the flesh without making it rock-hard. Even today, fish are almost always displayed on ice at markets around the world. In the nineteenth century, natural or machine-made ice allowed for the transportation of fish by rail hundreds of miles away from the places where fishermen first caught them. When New York's Fulton Fish Market opened in 1882, it sold fish shipped in by rail from as far away as Quebec.[81]

Fish lost a significant amount of value when frozen since the eyes tended to dry up and the skin became hard and loose on the flesh. It is still very difficult to preserve fish despite the availability of modern refrigeration, since the bacteria that make it decay are resistant to any temperature above freezing.[82] Refrigeration or freezing can prevent the spoilage of fish, but it still cannot preserve taste. Before the introduction of portable ice machines around 1920, fishing boats were limited in the distance they could travel by the time it took for the ice on board to melt. In hot weather, or if the boat were delayed for some reason, all the fish on board could spoil.[83] Starting in the 1970s, the Japanese began the fast-freezing of high-quality salmon on barges at sea, but this was practical only for large-scale fishing, and it significantly increases the price that consumers have to pay to eat it.[84] Today, ice machines have become small and cheap, making it possible for fishing boats to travel farther than ever before since they can just make their own ice as needed.[85] "Commercial fishing simply wouldn't be possible without ice," writes Sebastian Junger in *The Perfect Storm*. "Without diesel engines, maybe; without loran, weather faxes, or hydraulic winches; but not without ice."[86] Even flash-frozen fish are still thawed and displayed on ice in stores because customers associate ice with freshness.[87]

Oysters too have a long history of preservation on ice. One of the first trains to take advantage of the new transcontinental railroad in 1869 carried live oysters from New York Harbor to San Francisco Bay. By 1870, shipments of spat, baby oysters for cultivation in San Francisco Bay, began.[88] To go across country, the oysters were packed tightly in barrels and shipped on refrigerated railcars. Just as was the case with meat, shippers iced the cars and then re-iced them at stops along the way. The entire trip across country took between twelve and twenty days. While the protected environment inside the shell could keep most oysters alive for twelve days, if the trip took much longer, the mortality rate of seed oysters escalated sharply. Once the shipment arrived, oystermen dropped the oysters in promising waters near the coast for cultivation. Some people claimed that the Atlantic oyster actually tasted better when it grew in Pacific waters. Unfortunately, the Atlantic oysters that grew quickly in colder Pacific waters could not spawn. This necessitated continual shipments of seed oysters to keep this industry alive, at least until shipments of Japanese oysters (later renamed "Pacific" oysters) took hold later in the century.[89]

Other countries did not take advantage of refrigerated transport to the same extent that the United States did. Many did not need to. "The narrow width of England is responsible for the slow development of refrigerated transport," explained the refrigerating engineer A. E. Miller in 1948. He made special note of fresh fish being shipped inland from ports all around the country without a need for mechanical refrigeration. Geography cannot explain the slow development of refrigerated transport in the rest of that continent.[90] In 1910, America had 135,000 refrigerator cars compared with only 1,085 refrigerator cars in all of continental Europe. Most food consumed in the entire continent came from no farther than fifty miles away.[91] Germany had many of the refrigerator cars in Europe but used them most often to ship beer and butter rather than meat.[92] Russia had many of the rest, since the long distances between population centers in that country made investments in railroads an economic necessity. By 1910, a brisk trade existed for Russian meat sent to Germany for consumption. In 1912, a German refrigeration executive reported to his colleagues that "the meat smells fresh on arrival in Berlin and is in perfect condition," but he also stated that its appearance was not as good as meat coming from inside Germany. Discoloration generally set in despite the fact that the trip was only a day and a half (as opposed to the multiday journeys common in the U.S. market). The short duration made re-icing the cars unnecessary, though re-icing might have prevented any spoilage at all. Russia, which had much longer distances to cover than anywhere else in Europe, never had enough refrigerated cars to keep up with the demand.[93] The United States undoubtedly had the best land-based cold chain in the world. Besides meat, transcontinental shipments of fruits and vegetables depended upon ice for preservation, since it took many years to perfect mobile mechanical refrigeration.

From Farm to Plate

"In the cold parts of the country, don't you think people get to wanting perishable things in the winter—like peas and lettuce and cauliflower?" asks Adam

Trask, the protagonist of John Steinbeck's 1952 masterpiece *East of Eden*. "In a big part of the country they don't have those things for months and months. And right here in the Salinas Valley we can raise them all the year round." In much the same way that Thomas Mort wanted to carry surplus Australian lamb to a market he knew would crave it, Trask tries to ship his lettuce to New York. His inspiration for this idea is his icebox. "You know," he tells auto dealer Will Hamilton, "if you chop ice fine and lay a head of lettuce in it and wrap it in waxed paper, it will keep three weeks and come out fresh and good." Unfortunately, he faces two problems in making this dream a reality. The first is cost. Since he buys a local ice plant to make this possible, he has high start-up costs. When he faces his second problem, delays caused by weather and technical problems, the need to sell the "six carloads of horrible slop" that arrived in New York comes close to bankrupting him.[94]

Steinbeck captures the ups and downs of early produce shipping beautifully. Initial experiments with fruit and vegetable transport by rail did sometimes fail miserably. As late as 1919, U.S. Department of Agriculture inspectors found that decay had set into 11.8 percent of fruit and vegetable shipments into Chicago, Cincinnati, and Pittsburgh (and this did not count the perishables that still arrived via cars that were ventilated rather than refrigerated). Decay inevitably got worse after the produce left the cars.[95] On the other hand, Steinbeck's timing for Trask's misadventure was an act of extreme poetic license. Steinbeck sets Trask's failure, "a sensation in a year of sensations," in 1915.[96] In fact, producers first developed and shipped iceberg lettuce in 1903 specifically to prevent what happened to Adam Trask from happening in real life.[97] Include the shipment of perishable food by boat from California, and you have to go back further in time to find its beginnings. Transporting fruits and vegetables successfully, however, particularly with only ice as the source of refrigeration, took years to perfect.

For most of the late nineteenth century, meat was the only perishable food shipped by rail. During the 1890s, the first large-scale experiments shipping other produce began. Total refrigerated freight volume nearly doubled between 1901 and 1911. That growth continued until World War I, which disrupted normal operations. After the war, refrigerated rail transport volume continued to grow.[98] In California, refrigerated transport became a necessary prerequisite for the development of modern mass-production agriculture. Refrigerated transport made it possible for California to become the nation's primary source for fruits and vegetables because its products could reach the East Coast without spoiling. Early on, this trade proved so lucrative that 75 percent of refrigerator cars returned to California empty because the huge demand for them there meant that railroads did not want them waiting around for reloading. Westbound traffic existed for the services of these cars, but shipping California produce east simply made much more money.[99] By 1915, shippers had developed the collapsible ice bunker so that they could save space for nonrefrigerated goods on the return trip.[100]

Fruit cars were ordinary railcars with extra ventilation. They had very simple designs—often just freight cars with two side doors. One was for packing while the other, barred one was left open for ventilation. Fruit cars also had rooftop screens for more ventilation. They preceded refrigerated railcars, coming just about as early as iced cars did for meat. Fruit growers used these cars to ship fruits and vegetables by rail for the first time during the 1860s.[101] Since they had no ice, reported the *Sacramento Bee* in 1886, "one very hot day in transit will often ruin the greater part of a shipment."[102] The cargo also had to survive the cold during winter, which explains why many came equipped with stoves or heaters.[103] Despite these limitations, the *Omaha Bee* reported in 1874 that companies made two dollars from fruit shipped in fruit cars for every dollar invested.[104] By 1886, the Santa Fe Railroad sent ten to eleven fruit cars a day east during shipping season.[105] This volume made that profit possible despite the more than occasional spoilage or even the failure of entire shipments.

It took a long time for fruit cars to give way entirely to ice-packed refrigerated railcars. Despite the numerous failings of fruit cars, many railroads ran them well into the 1890s. The California produce industry depended upon fruit cars until the mechanical production of ice became common.[106] The peak years for fruit cars on the Santa Fe came between 1894 and 1896. Despite improvements in refrigerator car technology, many fruit cars remained in operation after the turn of the twentieth century.[107] As time passed, technological innovations and improved practices related to refrigerator cars made them more effective and therefore worth the added expense. For example, the way fruit was packed affected the success of the operation. Unless there was space between each fruit, air would not circulate and the product could get damaged.[108] More importantly, the pomologist G. Harold Powell developed the technique known as precooling in 1904. This simply meant keeping produce in an ice-cooled room before loading it on a refrigerator car or precooling the car itself by blowing air through the cargo before loading the ice.[109] This technique lowered ice consumption and meant that less cargo spoiled over the course of the journey. Later on, shippers used mechanical refrigeration for precooling even though ice remained the only form of refrigeration available en route. By making refrigerated transport more successful, precooling further increased total volume and made further specialization possible at both ends of the railway line.

Because of the high cost of infrastructure, most notably expensive cars and ice plants, only large companies could take advantage of the economies of scale needed to make California agriculture pay. At first, meat packers dominated fruit and vegetable shipments because of their previous experience with refrigerated transport. Since they had already built up their distribution networks in order to ship meat from place to place, handling butter, eggs, poultry, and fruit too made it possible for them to get the most from their investment.[110] The efficiencies discovered while handling meat allowed the packers to pressure the railroads into letting them handle most produce trains themselves. In 1905, one paper accused the Beef Trust of sending the price of fruits and vegetables up from between 10 and 25 percent.[111] Such arguments ignored the fact that refrigerated transport increased the supply of perishables, in some places during some months making fruits and vegetables available when the supply would not have otherwise existed. Still, the government put increasing pressure on the packers to give up their chokehold on the produce shipping business.

As a result, railroads became increasingly involved in these kinds of shipments. A milestone on the road to the removal of packing firms from this business was the formation of the Pacific Fruit Express (PFE) in 1906. Jointly owned by the Union Pacific and Southern Pacific railroads, PFE quickly dominated the perishable food transport business out of California. While the Armour Car Lines and a few other private companies sent some California produce eastward before the turn of the twentieth century, only with the formation of PFE was there a sufficient increase in volume to make California agriculture big enough to begin to feed the rest of America in any significant way. PFE's initial order of 6,000 refrigerator cars (or "reefers," as they came to be known) immediately increased the volume and quality of California produce available in eastern markets. In 1908, the firm shipped out 40,000 carloads.[112] Most refrigerated transport space passed from private car lines (usually owned by packers like Armour) to railroad ownership around 1910.[113]

Cantaloupes demonstrate the scale of PFE's operations well. PFE deployed a thousand people for the cantaloupe harvest, five hundred just to ice the reefers brought in from all over the Southern Pacific's vast system. Since temperatures during the fruit's six-week harvest season in the Imperial Valley could go as high as 130 degrees Fahrenheit, the railroad required enormous amounts of ice from the outset. Railroad workers iced each car as many as nine times during its trip east. By 1927, cantaloupes from the Imperial Valley appeared on grocery store shelves around Chicago just a week after being harvested. With this distribution system in place, cantaloupe production soared. From only 237 carloads heading east in 1905, the Imperial Valley of California was the departure point for 21,500 carloads of cantaloupe in 1931. This constituted half the cantaloupe production of the entire United States.[114] Multiply those kinds of numbers times the number of perishable crops grown in California and the role of ice in making California the center of American agriculture becomes obvious. Without ice, the state could never have developed as large an agricultural industry as it did during the early twentieth century.

Considering that a single carload of cantaloupe needed 10,500 pound of ice for its first packing and 7,500 pounds for each re-icing, it makes sense that the railroad needed the biggest ice plant in the world. Like Swift before it, PFE quickly acquired or built a large network of ice-manufacturing plants all around the country to service its trains.[115] As the company grew, its ice-making facilities also expanded. It went from using outside suppliers for its ice to manufacturing ice itself right where it was needed, thereby shortening the length of the cold chain for the ice needed for perishable food. By 1924, the firm operated many of the largest and most technologically sophisticated ice plants on the West Coast and at other places along its lines.[116]

During the twenties, some California produce eventually came to rest in the first electric household refrigerators, but ice still played a crucial role in the land-based transport of perishable food for another thirty years. Only abattoirs and storage facilities used mechanical refrigeration because of the size of these machines. They did not fit any part of the cold chain except for ships. Ice (first natural, then manufactured) stayed so cheap and readily available that there was no financial incentive to develop mechanical refrigeration for railways in the early twentieth century. Because of technological difficulties and cost, a commercially viable railcar that used mechanical refrigeration to keep its cargo cold did not appear until the 1950s.[117] Despite widespread interest, the majority of refrigerated railcars in the United States still depended upon ice in 1977, probably because of the cost of this new technology as compared to just using ice.[118]

The continuing importance of refrigerated transport to the American diet suggests the continuing role of geography in food chains of all kinds. The obstacles presented by topography and climate gradually receded as the technology of both refrigeration and transportation improved. Producers ship food out of places like California because it is produced most efficiently and in large batches there. Grocers buy food shipped in from places like California when local producers cannot produce it more efficiently or during winter. Travel does nothing to improve its taste. "Unfortunately, the full richness of flavor is found only in the fruit that matures upon the vine, or tree," complained the journal *Ice* about the practice of shipping green fruit by rail in 1909, "and thus New York, while eating ripe California fruit, never has known how good California fruit ought rightfully to be."[119] Most New Yorkers *still* do not know how good California fruit ought to be because no matter how advanced refrigerated transport has become, these technologies cannot overcome the physical limitations of the fruit.

As successful as refrigerated transport has become, an effective cold chain also must protect perishable food against the passage of time. Without the ability to successfully store such perishable food products, even a successful shipment along the cold chain could spoil upon arrival if it had to wait around for a ship or another railroad car. That explains why advances in cold storage technology quickly followed those in refrigerated transportation. The obvious need for this technology should have led both producers and consumers to embrace it, but early problems with cold storage called the usefulness of the entire cold chain into question. This led critics to attack cold storage as a menace to America's physical and economic health. Despite resistance to cold storage, the United States nonetheless took to this technology faster than any other country in the world. By the 1920s, it became yet another part of the cold chain for which the rest of the world fell behind.

CHAPTER 5

The Pleasures and Perils of Cold Storage

Dr. Harvey Washington Wiley ran the Bureau of Chemistry at the U.S. Department of Agriculture from 1882 to 1912. In that capacity, he helped convince Congress to pass the landmark Pure Food and Drug Act of 1906. Speaking before a class in sanitary science at Cornell University sometime after the passage of that law, he described his attitude toward the cold storage industry. "I don't object to cold storage," Wiley explained. "Cold storage is a glorious gift to man. It is the best method in the world of getting food to market in a fresh state." Wiley had run experiments demonstrating that common chemical preservatives hurt people's health. Cold storage obviously seemed like a better choice, but he did not give the purveyors of this technology his complete support. "What I object to, is not its use, but its abuse," he told the class. "It is a means of restricting the supply of foods at one time, in order that the price may be raised at another time." Wiley also objected to the effects of cold storage on the freshness of some products. "There are certain things that ought never to be kept," he told the students. "Some of these are butter, milk, cream, eggs, poultry and fish, all staple articles of diet, and every one of them should be eaten fresh."[1] Unfortunately, the cold storage industry depended upon precisely these products for most of its business. Taken together, these two objections—the effects of cold storage on both prices and food quality—made this one of the most important (yet now completely forgotten) controversies of the Progressive Era. Congress held repeated hearings on whether and how to regulate this industry. (The fact that those regulations never passed might explain why nobody remembers this issue anymore.) State legislatures eventually proposed (and some passed) their own regulations too.

In order to resolve the controversy surrounding cold storage, Wiley conducted a series of experiments designed to determine the effects of cold storage on food. After those experiments ended in 1910, Wiley informed the U.S. Senate of his conclusion: consumers could safely eat food kept in cold storage. "Storage is a legitimate exercise of the activities of the food distribution," he explained, "when it is practiced for the purpose of bringing a fresh article to the market in the best possible condition." While Wiley recognized that this process might increase the price of foods kept this way, he also recognized the willingness of the consumer "to pay this increased cost because of the increased value of the food product which he secures."[2] Because of his clout with the American public, Wiley's support did much to ward off efforts to regulate the cold storage industry at the federal level.[3]

Wiley's conclusions about the safety of foods kept in cold storage hinged on his implicit assumption that all cold storage facilities performed their function the same way. In other words, he tested the effect of cold storage on different foods rather than the equipment itself. In truth, cold storage facilities varied greatly in quality at the time Wiley conducted these tests. Before World War I, the industry had not yet acquired the knowledge or even the technology to keep perishable food fresh with great reliability. Whatever ill effects cold storage had upon the healthfulness of food resulted from technological failures, not the cold storage process itself. Since most of the technological problems associated with cold storage dealt with the inability of the technology to deliver the precise temperatures needed to preserve any particular food best, the learning curve that came with experience alone did much to solve such problems. The convenience of buying food out of season more than made up for the economic costs of cold storage once the industry perfected this technology. When cold storage became both effective and commonplace, the controversy surrounding it quickly subsided.

Some concerns about the taste of goods kept in cold storage remained. As with perishable food that travels from far away, the taste of any edible good kept in cold storage for a significant length of time inevitably degrades. As the distance between the point of production and the point of consumption for food products of all kinds began to expand, consumers wanted to know more about what happened to what they ate over the course of its journey. Labeling cold storage goods would have helped to solve that problem. This kind of regulation failed to materialize because Americans eventually adapted to the taste of food preserved through cold storage. Convenience and variety eventually trumped taste in their order of priorities. Once cold storage became commonplace, it quickly faded into the background in the United States. Cold storage goods did not play as big a role in the food systems of other countries because consumers elsewhere resisted any food preserved this way. Besides, since the United States had a much more sophisticated cold chain than those in any other country, the use of cold storage never became an important issue elsewhere until after World War I.

The first successful large-scale efforts to preserve food over time for later consumption out of season had a considerable impact on the American diet. Cold storage not only changed the daily menu of the average citizen the same way that refrigerated transport did, but it also changed people's expectations. "The effects of the [cold storage] industry have become thoroughly intrenched [*sic*] in civilized man's habits of living," wrote the *New York Tribune* in 1910. "He expects to eat eggs all the year round, while his grandmother used them in the winter only for cake baking. Likewise, apples in May or June are taken as a matter of course, whereas preceding generations did without."[4] As the price of perishable foods kept by cold storage dropped, luxury foods became necessities. Cold storage remains a vital part of the cold chain today, but the anxieties that cold storage exposed early on demonstrate the difficulties that American consumers had adjusting to such an important, far-reaching technology.

The Structure of the Cold Storage Industry

Before cold storage came along, everyone recognized the value of any technology that could preserve perishable foods over time. Many people in nineteenth-century America looked at the enormous waste of food resources and hoped for a solution. "In spite of universal care and forethought," wrote the newspaper editor Horace Greeley in 1864, "millions' worth of food are destroyed daily through its action, and innumerable buffaloes and other cattle are slaughtered for their hides, and the carcasses left to rot where they fell, because the conditions necessary to their preservation for food do not there exist."[5] While Greeley might not have known about it when he wrote those words, the first experiments with cold storage in the United States had already begun. Like so many other aspects of the cold chain, it took decades for these experiments to succeed, but the clear utility in the service that cold storage supplied convinced people to accept its many initial shortcomings. Most of the early users of cold storage were producers who could conceal its use as long as the passage of time did not unduly affect the taste of their food.

The first cold storage facilities used ice, but most quickly came to depend upon ammonia compression refrigeration once that technology became reliable and affordable. Since ammonia could taint food, such cold stores used their machinery to chill brine that circulated in the food storage locations of the warehouse. All the earliest efforts along these lines involved food production facilities. Starting in 1858, Chicago pork packers used ice to pack pork during the summer and keep it onsite, something they had previously done only in the winter. Initially, cold storage did not always work as intended because it was impossible to precisely control temperature using just ice as the source of cold or to control for humidity.[6] The profitable example of pork refrigeration led to the development of refrigerated storage for slaughtered cattle too. Cold storage both limited the amount of decay that occurred after death and allowed packers to store beef to time its entry into the market. Because of cold storage and refrigerated transport, Chicago dressed beef changed America's eating habits starting in the 1870s.[7] The gradual introduction of mechanical refrigeration into the packing process made it possible to chill meat faster and to control the temperature better, thereby improving the quality of the product that consumers ate.[8]

The earlier ice-refrigerated cold storage warehouses generally had two stories and depended upon an ice and salt mixture as the source of the cold. The first floor usually held perishable products, and the owners would pack the second floors with ice. Openings in the floor allowed the cold air to descend and chill the contents of the bottom story.[9] An early English design for an ice-refrigerated cold storage unit divided the building into eleven arches separated by airtight doors. One compartment contained the ice supply for the building, while the cold air that came through earthenware pipes connecting the ice compartment to the rest of the building regulated the temperature. A fan connected to a two-horsepower engine blew the cold air through the pipes. The food sat on shelves on the upper stories of the building. The designers believed that the entire building could stay under 40 degrees Fahrenheit; however, any kind of precise temperature control under such arrangements was impossible.[10]

The first cold storage facility that used mechanical refrigeration appeared in Boston in 1881.[11] These kinds of facilities also tended to be multistory buildings. The design reserved the first floor for loading and unloading onto wagons or, later, trucks. The goods to be stored went by elevator to the upper floors, where, in theory, different sections stayed at the ideal temperature needed to keep a wide range of products.[12] Such cold storage warehouses could rise as tall as fifteen stories, and some larger warehouses had facilities to run railway cars right into the building.[13] Most warehouses had brick exteriors and concrete floors to make them easier to clean.[14] Many used mechanical refrigeration to chill brine that circulated throughout the facility, since ammonia would taint food if it ever leaked.[15] Warehousemen carried special insurance policies that limited their liability if their equipment failed and any of the food in the building spoiled as a result. The owners of cold storage warehouses with mechanical refrigeration also had to pay two to three times more in fire insurance than did the owners of ordinary warehouses, because of the risk of ammonia explosion, but that did not make cold storage a controversial technology.[16] The perceived effects that this technology had on the perishable food that it preserved quickly became a much bigger problem.

"Cold Storage Evils"

Food systems depend upon trust. Food is unlike other products in that people consume it by putting it into their bodies. This means it affects their health, not just their wallets. Already concerned about the adulteration of food, Americans also worried about the effects of cold storage upon what they ate.[17] Because of these fears, many hysterical headlines concerning the health effects of cold storage appeared during the early years of the twentieth century. An industry representative singled this one from 1911 out for special attention in the pages of *Ice and Refrigeration*:

> Cold Storage Evils—Thousands of Tons of Food Unfit to Eat Foisted on Public By Freezer Owners—Saps Nations [*sic*] Health—Bad Eggs, Poisoned Poultry, Deadly Fish, Unwholesome Butter, and Decaying Vegetables Kept to Get Benefit of High Prices—Science of Keeping Eatables in Good Condition Not Known to Storage Men—A National Disgrace.[18]

Like Wiley's speech to those Cornell students, this clear example of yellow journalism also serves as a good summary of public concerns about the cold storage industry in the pre–World War I years. The effects of cold storage on both health and food prices worried consumers. As a result of these concerns, the cold storage industry had to deal with multiple attempts to regulate its operation until technological improvements and changing political circumstances made its services part of the accepted practice of American food chains.

Press coverage during this era strongly implied that problems with the wholesomeness of cold storage foods arose as a result of the industry's lust for profits. This critique echoed concerns about the health risks of eating food maintained by new kinds of chemical preservatives.[19] While the press suggested greed as the underlying cause here, the industry's ignorance of how best to employ its own technology explains this situation better. For example, warehouse operators often kept fish and meat out in the open, where they often tainted other products and eventually spoiled. Many operators actually feared keeping anything at too cold a temperature.[20] Those in the cold storage industry had every interest in keeping their products fresh, but they lacked the technical knowledge necessary to achieve that goal. Unfortunately, it took years for a suspicious public to cut them some slack.

Cold storage operators used those years to work out the science associated with their business, which helped them learn how to operate their facilities effectively. Until then, cold storage operators ran their untested technology without an instruction book. This inadequate knowledge created an environment ripe for lawsuits as people storing goods blamed the operators of cold storage facilities for less-than-palatable merchandise and the operators blamed the dealers.[21] Regardless of fault, such situations severely damaged the reputation of the cold storage industry in America and around the world. Many consumers associated spoiled food with its time in cold storage whether this technology deserved the blame or not.

The technology deserved a great deal of blame in the years before operators learned to keep different foods at different temperatures. Every product kept in cold storage stayed freshest within its unique range of acceptable temperatures. Otherwise, it spoiled before its time. Yet no universally accepted exact temperatures for each food product commonly found in cold storage existed until the 1920s.[22] Every cold storage publication offered different suggested temperatures for every perishable commodity. In truth, ideal temperatures actually depended upon the conditions where the goods originated, so no chart could capture every nuance required for successful cold storage. To make matters worse, as long as warehouses used ice for refrigeration, the ability to keep anything at a steady temperature remained in doubt. Melting ice put moisture in the atmosphere, and moisture promoted mold. Ironically, the potential for problems with ice increased during cold weather. If the weather stayed cold enough so that the ice would not melt, the impurities that collected on the ice would evaporate and circulate onto the food.[23] Despite such problems, the amount of food kept in warehouses cooled by ice far exceeded the amount cooled by mechanical refrigeration until 1905.[24] This alone explains the failings of cold storage up to that date.

"Of course," wrote the cold storage expert Madison Cooper in 1900, "there are many in the business, even now, who do not know anything about cold storage, except temperature."[25] Successful cold storage also depended upon air circulation and ventilation, but most of all, the condition of goods in cold storage depended upon humidity.[26] And there was a fine line in this regard. "When there is too much moisture in the air, goods grow moldy quickly," explained the refrigerating engineer Milton Arrowood in 1917. "When there is too little moisture in the air, they dry out."[27] The degree of humidity in any given room also depended upon the amount and kind of goods stored there, since food can release both moisture and heat.[28] Even if temperature and humidity stayed steady, the possibility of cross contamination by scent remained. This explains why warehouses had to learn to keep fish or onions separate from all other food products.[29] Likewise, less smelly products, like apples and potatoes, kept together in bulk and lying loose on the floor could leave the warehouse tasting like one another.[30] Products like pears when kept alone could ferment and give off gasses if stored improperly. Eggs needed open space around them because they can give off a lot of moisture and are quite sensitive to changes in humidity.[31] Although this made ventilation a necessity, a refrigerating engineer took to the pages of *Ice and Refrigeration* in 1898 to complain that "not one house in ten" had "any kind of decent ventilation."[32] Poor ventilation or lack of humidity control also led to a lot of spoiled food, which in turn created concerns about the health of consuming perishables kept in cold storage.

Spoiled food could also come from repeatedly thawing and rechilling the product. Merchants who sold frozen or cold storage goods often defrosted these goods in order to fool consumers into mistaking them for fresh.[33] If consumers could not easily tell the difference in the store, they could when they tasted their food after they got home. That hurt the reputation of the process. In the case of meat, consumers had an easier time recognizing products that had been kept in cold storage. The effect of freezing on meat tissues meant that meat would not keep as long once thawed. At a time when refrigerated display cases had not yet caught on, this practice invariably damaged the meat more than the cold storage process itself.[34] Even when thawing and rechilling did not necessarily make the meat unhealthful, it still made the meat less appetizing and less appealing to the eye. Imagine a cut of meat thawing and being refrozen day after day, and it is easy to see why cold storage meats had a bad reputation. "Some merchants seem to think the practice of cold storage is some kind of dishonorable performance and indignantly deny that they practice it," explained Wiley in 1905.[35]

Ironically, the bad reputation of cold storage foods explains why they froze and thawed meat at all. Hiding time in cold storage because of its bad reputation only made its reputation worse.

Some food spoiled before it ever arrived in cold storage. Despite the many food processors that built cold stores to hold surplus product, many holes remained in the American cold chain during this era. Perishable food often sat unrefrigerated as it waited transportation or sale. "Cold storage, one of the greatest developments of our time," wrote the home economics pioneer Mary Hinman Abel in 1921, "has hardly been incorporated on a large scale into our system of food distribution. . . . Better refrigeration will stop the waste by spoilage." The failure of American cities to develop terminal markets that could refrigerate perishables in the course of their distribution frustrated her in particular.[36] No city had a more advanced market system than New York, yet it had gaps in its cold chain well into the 1920s. "When products reach the market," complained New York City's commissioner of public markets in his 1923 annual report, "they are almost universally removed from the cooled car to a platform or warehouse which often maintains high atmospheric temperatures. Eggs, butter, poultry and vegetables stand for hours in the summer sun or beneath a thin wooden roof, and are then carted by unprotected trucks through the hot city streets to markets which are not provided with mechanical refrigeration and wherein even the icebox is of antiquated construction."[37] As a result of such criticism, the city did build cold stores for its public markets by the end of that decade. This helped, but cold stores at markets could only keep food for so long until new merchandise arrived.

Similarly, anything that was already damaged when it went into cold storage remained damaged upon coming out. Reporting to Congress on the state of cold storage, Harvey Wiley said: "We found that there was no uniform system for inspecting products that went into cold storage; almost anything that was brought to a cold storage man—he didn't own the goods; he was paid for keeping them; it wasn't his business to inquire into their nature or character. Practically anything which was not flagrantly decayed, and so forth, would be accepted under such circumstances, and properly so, as far as I can see."[38] Warehousemen often stamped the boxes they held with the wording, "Conditions unknown. Stored at owner's risk."[39] Already bruised fruit cannot heal while in cold storage. As the rest of the fruit ripens, the point of the bruise will only blacken.[40] A bad egg could become bad at any point from inside the hen to the cold storage warehouse, and nobody could tell when exactly it spoiled. Yet this did not stop consumers from blaming cold storage for the problem, even if it was a time of year when no eggs ever went into cold storage.[41]

Had customers understood more about the entire cold chain, they never would have developed a prejudice against this method of preservation. Despite the technological problems, the failures of early cold storage technology did not justify most of the public's concerns. For example, many critics of the industry during its first years feared that producers kept food in cold storage too long.[42] In reality, few producers kept meat over three months, as every kind of meat tended to go in and out of cold storage in three-month cycles.[43] Fruit had a longer cycle, but the limitations of keeping it long term prevented anyone from storing it too long. "Nobody ever holds apples over twelve months," explained Senator James Wadsworth of New York while discussing a federal cold storage regulation bill in 1920. "A man would be a perfect idiot to do it, because he would be swamped by the incoming fresh crop."[44] In other words, a year-old apple would have still have been edible but not as desirable as a fresh one, and the newer apples would have sold out first, so producers never kept them that long. Perishables decay even in perfect cold storage, just very slowly. To see food in cold storage as part of a dichotomy between fresh or spoiled confuses things. Cold storage goods ought to be considered on a sliding scale. The term "fresh," in other words, is relative.

Whether a consumer risked buying a cold storage egg rather than a fresh one usually depended upon its price. Some critics suggested that cold storage, like refrigerated transport, led to a large increase in food prices during the first decades of the twentieth century. "The cold storage business occupies a strategic position in the economic field," wrote the muckraking journalist William Leavitt Stoddard in 1914, "It thwarts the 'natural law' of supply and demand by collaring the supply and shelling it out when the demand comes in such a fashion as to secure the greatest amount of money from the neediest folks at their hungriest moments."[45] This kind of argument had precedents that dated back to the earliest stages of capitalism. Engrossing, forestalling, and regrating were all felonies under English law. The first of these terms meant buying up large quantities of perishable food, the second meant keeping them off the market to fetch a better price, and the third meant selling what you had hoarded.[46] The cold storage industry depended upon doing all three of these things together, but overall, the effects of these practices helped consumers more than they hurt them.

Producers almost always kept cold storage goods in warehouses that they did not own. The warehouse owners seldom had any refrigeration business.[47] Food producers contract with these warehousemen for space. An observer noted in 1911 that there were a thousand cold storage plants in the United States, each with "scores and scores of customers, and each of these customers is in keen competition with all his fellows. To combine them would be about as difficult as to combine the farmers."[48] Too many people influenced the day-to-day operation of the industry to even dream of setting the storage rate on anything. In fact, anyone wanting to store a large quantity of goods in a city with multiple cold storage houses could shop around, pitting warehouse against warehouse in order to get the best rates.[49]

Cold storage did affect consumers' food prices somewhat, but it could either raise or lower them, depending upon the circumstances and the timeframe considered. As the Senate Committee on Manufactures noted in 1911, "The effect of cold storage on prices is in general to make them steadier, preventing extreme fluctuations either upward or downward."[50] While producers certainly hoped to get higher prices for products they kept in cold storage, the ability to store food also increased supply, which forced prices down if stored and fresh goods entered the market at the same time. Overall, the cold storage industry served as a balance between supply and demand, making it possible to bring extra supply on the market when demand existed and to take supply away when demand did not exist.[51] Cold storage also improved the supply of all perishable foods by preventing waste. Any short-term price spikes when a producer withheld a particular good from the market were offset by the long-term effects of the increased supply resulting from its preservation. A modern analysis of butter prices during the 1890s (a commodity for which excellent records exist), suggests that cold storage caused summer prices to rise between 3 and 17 percent (when some of the supply was set aside for cold storage) and winter prices to fall between 3 and 12 percent (when butter made in the warm months came out of storage).[52] Complaints about the summer price rise drowned out praise for the winter price decline for a time, but the advantages of cold storage eventually permeated the public consciousness thanks to the onset of World War I.

"Our Best Friend"

"Was it just a passing headline fancy that charged the cold storage system to be a criminal conspiracy to raise prices on essential food products?" asked the political scientist Walter E. Clark in 1915.[53] As a matter of fact, it was a passing headline fancy. In 1913, the home economics pioneer Mary Pennington, then affiliated with the Food Research Laboratory at the U.S. Department of Agriculture, explained why attitudes towards the price of cold storage food could change so quickly: "If the consumer can be taught that at certain seasons of the year his demand for certain commodities can only be supported by way of the cold storage warehouse and that he must either accept cold storage products or do without, he will then very promptly develop common sense methods for the utilization of these products."[54] This reason alone explains why the public attitude towards cold storage improved immediately before and during World War I.[55] The acceptance and even approval of cold storage in America has lasted to this day.

Improvements in the effectiveness of cold storage technology did much to make this change in attitude possible. A large increase in the number of cold storage houses that used mechanical refrigeration made the most significant difference during these years. By 1914, only seven warehouses in the whole country still used ice as their source of cold.[56] Just as importantly, the knowledge deficit that had once existed in the cold industry gradually disappeared. In the early days of the industry, warehousemen independently engaged in a series of expensive experiments to figure out the best temperatures for the goods that they most commonly kept.[57] When such knowledge became widely distributed, the effectiveness of the entire industry's technology improved. These efforts culminated in the 1922 founding of a lab entirely devoted to cold storage experimentation, the Marble Experimental and Research Laboratory in Canton, Pennsylvania. Mary Pennington served as the director of research there.[58] Once science had helped to improve the quality of products kept in cold storage, they became practically indistinguishable from fresh ones. This in turn increased supplies, which would have forced prices down, thereby making cold storage significantly more popular.

The reason that America's entrance into World War I improved public attitudes toward cold storage revolved around fear of wartime food shortages. Since Americans wanted to save food for both their own soldiers and starving European allies who could no longer feed themselves, cold storage became a much-touted part of the overall war effort. President Woodrow Wilson created the U.S. Food Administration to increase production, discourage waste, and conserve America's food supply so that it could help Europe feed itself during the crisis. The Food Administration's leader, future president Herbert Hoover, refused to ration food at home. Instead, he encouraged Americans to conserve it voluntarily and bolstered this effort with an elaborate propaganda campaign. To reject cold storage food under these circumstances would have seemed unpatriotic. Food Administration rules imposed on the cold storage industry made what the industry did easier for consumers to accept too. For example, throughout the war merchants could not sell poultry, eggs, butter, or fish as fresh when kept in cold storage for more than thirty days. To prevent speculation in cold storage goods, the Food Administration limited the amount of loans warehouses could give small food producers who were storing perishables with them (a favorite way to keep those warehouses full), thereby making it much harder to time the market for any perishable product in order to maximize revenue. "The purpose of the regulations," explained the agency when it announced them, was "bringing the operations of all storage concerns out into the open," which in turn quieted consumer fears.[59] With the government so aggressively involved in this industry, lingering trust issues disappeared.

Seeing an opportunity, the cold storage industry used the war to improve its public standing. It immediately volunteered to cooperate with Hoover to whatever extent he desired, recognizing that it played the role of a "semi-public utility."[60] Throughout the war, the cold storage industry cooperated with the Food Administration in many ways ranging from the licensing of their facilities to reporting the amounts of all the various types of food that warehouses stored. It helped that regulations placed upon the industry were largely written by a cold storage executive who served in the Food Administration and wrote them in a way that worked to the industry's benefit.[61]

The cold storage industry's multifront public relations strategy proved very effective. For example, as car after car of "eggs, string beans, eggplant, lettuce, peppers and potatoes" that had been kept in cold storage since their harvest the previous year arrived in New York City during January 1918, produce prices dropped despite the demand for food generated by the war. As a result, the *New York Tribune* reported, "the despised, dreaded and suspicioned [*sic*] cold storage facilities are now our best friend."[62] When the war ended, all the extra cold storage space built for the government's war effort was turned over to private hands. Moreover, cold storage capacity increased sharply throughout the country in response to the greater public acceptance that the war had created.[63] In 1921, the meatpacker Swift & Company ran an advertisement in the *Saturday Evening Post* that revolved around a group of Eskimos visiting Edmonton, Alberta. "They had never before been south of the Arctic Circle," the ad said. What amazed them the most? Not the cars or the airplanes but the cold storage facilities: "The White Man didn't always have to hunt and fish when he wanted to eat!"[64] Racism aside, Swift's comfort in putting cold storage at the center of an advertisement demonstrates that public attitudes towards this practice had changed drastically. Instead of hiding its impact, Swift wanted to make sure that customers did not take that impact for granted.

This changing public attitude toward cold storage greatly improved the industry's fortunes in the political realm. In the 1910s, critics concerned about possible abuse had proposed regulating the cold storage industry. Initially, the industry put up blanket opposition to any kind of cold storage legislation. Testifying before the Senate Committee on Manufactures, the refrigerating engineer Victor Becker argued: "Not the cold-storage house, but the speculator, should be legislated against."[65] Food producers of all kinds even argued that cold storage food was actually an improvement on fresh. Despite the absurdity of this argument, the industry still beat back a 1910 bill from Senator Henry Cabot Lodge of Massachusetts that would have prohibited the sale of goods kept in cold storage for more than a year. Similar federal bills debated around this time also went down to defeat. As a result, until World War I the only successful legislation regulating the cold storage industry occurred at the state level.[66]

Between 1911, when criticism of cold storage peaked, and 1915, the number of states that passed laws regulating cold storage increased from five to eleven. The most common regulations included labeling laws, which forced cold storage operators to mark goods as they came into the warehouse so people would know the age of whatever they bought.[67] In 1920 the former director of the Food Administration's cold storage section, who had returned to his cold storage business, noted that about fifteen states had cold storage laws; the provisions of these laws were, he said, "reasonable and just and are based upon the so-called Uniform Cold Storage Law recommended by the Commissioners of Uniform State Laws," an impartial body which made its recommendations in 1914.[68] The industry actually grew to like many of these laws, since they affected unscrupulous operators most.

Improvements in technology and a better regulatory environment led to a huge increase in American cold storage capacity from the early twentieth century until the advent of the modern cold chain fifty years later (table 5.1). The 620 cold storage warehouses in existence in 1904 were scattered around the United States, mainly in cities. By 1916, that number grew to at least a thousand such warehouses. The greatest increase in cold storage capacity came between 1920 and 1925, after the technology had proven its value during the war. The

Table 5.1. Increase in cold storage capacity in the United States, 1904–1925

Year	No. of cold storage warehouses	Capacity (million cubic feet)
1904	620	102.5
1916	~1,000	250.0
1920	>1,000	273.4
1925	1,710	361.6

Sources: Oscar Edward Anderson, Jr., *Refrigeration in America* (Princeton: Princeton University Press, 1953), 128-29; *Ice and Refrigeration* 101 (July 1941): 5.

crowning achievement for the cold storage system came during World War II, when despite its being worked to more than capacity, only 0.1 percent of food purchased by the War Food Administration to feed the troops went bad.[69] Cold storage warehouses remain a vital element of the modern cold chain. Both long experience with the technology and now computerization make cold storage both efficient and effective.[70] Because other countries were behind the United States in terms of refrigeration use in the years before World War I, the general acceptance of cold storage obviously took longer elsewhere in the world.

Cold Storage around the World

In 1886, English plum growers faced a problem. The problem was not insect- or weather-related. It was, as the *London Daily News* explained, "a glut of fruit." The paper reported that all Continental plum growers got shut out of English markets that year because of the English surplus. In "some quarters of Kent that great quantities of plums are being allowed by the owners to rot upon the trees, as they find it does not pay them to gather them and send them to market."[71] In 1891, a correspondent to *The Standard* in London informed the paper that plums that year had become "unmarketable and rotting on the trees" for the same reason.[72] This situation illustrated the law of supply and demand: when a bumper crop of any agricultural product comes in at the same time, the prices will inevitably go down if the entire increased supply of that product is brought to market all at once. Yet English plum producers could not spread out the introduction of their fruit to market over time; the plums they held onto would have rotted because they had no way to preserve them. They let their plums rot on the trees because prices were so low that it was not worth investing any more resources in picking them. Once the glut of plums had passed through the system, prices for the fruit undoubtedly went back to normal in London and other markets.

American food distributors solved the problem faced by English plum growers by the development of cold storage. The United States and Great Britain grew very similar plums, and the most popular varieties could survive between eight to ten weeks in cold storage. But English plum producers had no cold stores to use for fruit.[73] Even though Americans kept plums only "for a few days at a time to tide over an overstocked market," cold storage still had a huge impact upon on the plum market.[74] Since plums tend to ripen in a very narrow window, this technology became essential for getting the price up and keeping plums a profitable fruit to produce in the United States. Despite earlier harvest disasters and experiments by the Kent County Council that demonstrated that plums could "keep for a long time in good condition while waiting for a favorable market," as late as 1906 British plum growers still did not use cold storage, even though their competitors on the Continent did.[75] Only English meat markets used cold storage.

By 1952, America had 135 square meters of cold storage space for every thousand people in the country, Great Britain had 50 square meters, and the French had just 25 square meters.[76] This situation arose because historically almost all of the growth of cold storage capacity in Europe was used for imported food near docks and not for internal transport of food. The first cold storage facilities at Smithfield, London's central meat market, were not built until 1889.[77] Despite elaborate operations there by 1894, with "tens of thousands of carcasses" arriving every week, the meat industry delayed the refrigeration of the entire meat import chain.[78] In 1897, vessels with refrigeration were repurposed solely for cold storage because London's land-based cold storage capacity had not grown fast enough.[79] By 1901, a few larger towns around the country had small cold stores for meat, but the industry needed more.[80] "The only thing that can check the progress of imported meats and give the English farmer a fair chance," one English refrigerating engineer complained in 1902, "is a universal system of cold storage in all our towns. At present the English farmer is severely and unfairly handicapped."[81] By then, even English plum growers probably agreed.

It was not just farmers and fruit growers who suffered as a result of Britain's dearth of cold storage facilities. Without a cold chain infrastructure, consumers outside of London could not take full advantage of the thousands of tons of chilled and frozen meat that arrived at Smithfield Market every day. Refrigerating equipment remained rare in London butcher shops as late as 1902, making the meat that much more difficult to preserve for sale.[82] Less preservation meant a smaller supply. A smaller supply meant generally higher prices. Countries without an infrastructure could still benefit from having some perishable foodstuffs sent to them, but consumption of chilled products would never take off in Great Britain the same way they did in the United States until companies all along the cold chain made similar investments.

Australia's modern cold chain for meat faced a similar problem with lack of infrastructure. Thomas Mort had established the first freezing works in the country at Bowenfels in the Blue Mountains outside of Sydney in the early 1870s.[83] Other concerns followed Mort's example after his death, trying to tap the same extraordinary potential of the almost limitless supply of sheep and cattle in the country's interior while avoiding the inherent pitfalls of shipping live animals. Unfortunately, not nearly enough cold storage existed in and around Sydney for either the local or the international dressed meat trade. "A proper chilled room to protect the meat should it not be sold on arrival is a *sine qua non*," wrote the American consul there in 1891. Instead, firms "had to take the price offered by the retail butchers or allow it to stink. They were, in fact, largely at the mercy of the trade; whereas with a proper chill-room the salesman could put the meat in and wait a fortnight if necessary for better prices." The abattoirs in that country did not have enough cold storage facilities nearby either, meaning that the dressed meat sometimes went into the freezing works already damaged.[84]

Argentina, on the other hand, went to great lengths to develop cold storage facilities in order to grow its meat export industry. The government not only waived export duties on fresh meat and subsidized the companies engaged in the trade, but it also waived duties on refrigerating equipment needed for the exporters to build cold stores. As it granted concessions to various firms for grazing lands, it required those firms to build additional cold storage facilities. Exporters built six new facilities in 1909 alone.[85] This allowed Argentine meat to significantly undersell the Australian product on the British market, thereby bringing the Australian export industry into a crisis by the early 1920s.[86] But Argentina's cold storage facilities did not make it easier for Argentines to eat fresh meat themselves. Because the cold chain in Argentina existed only for exports, the U.S. consul in Buenos Aires recommended consumer education as a way to create demand for American refrigeration equipment. Yet even his argument depended upon the improved safety of refrigerated food rather than the convenience of shopping less, a perceived advantage in America.[87]

Like Americans before World War I, French consumers had great suspicions about cold storage, formed on the bases of both safety and price. Unlike America, in France those suspicions delayed the development of the cold storage industry until after World War I. "The cold storage building of the Exposition of 1900 will be a revelation in Paris," wrote the journal *Cold Storage* the year before that event. "Nothing like it has ever existed there, and the possibilities of the artificial conservation of perishable products are very little known to the rank and file French, and generally speaking, to the people of the Continent."[88] In 1903, the entire country had only three cold storage warehouses.[89] People in France did not see this as a problem until 1911, when a government report blamed the high cost of food on the lack of cold storage facilities there.[90] The French did not trust refrigerated food, at least not at first. They liked being able to bargain with merchants for lower prices at the end of the day on stock which would otherwise spoil since there was no way to preserve it overnight. French consumers also feared that cold storage would prevent them from differentiating between goods that were preserved and those that were truly fresh. Even Paris's central wholesale market created rules designed to limit the use of cold storage. The French preferred to do their shopping at the market every day. Similarly, merchants preferred to keep their picky customers' trust rather than introduce a potentially disruptive technology.[91]

But after their initial opposition to cold storage, the French (like Americans) warmed significantly to the cold storage process. During World War I, "the success of Americans in handling perishables on a large scale in France opened the eyes of the French to the possibilities of the process," explained an American government official in May 1923.[92] Once the French learned to do cold storage well, the number of cold stores increased; by 1950, France had 882 cold storage facilities.[93] It seems only natural that Americans, the people who had the most experience with this technology, showed the French how to do it right.

By the middle of the twentieth century, cold storage had become a priority around the world because governments stockpiled perishable goods during the Cold War.[94] By then, not just equipment but the entire design of cold storage warehouses had vastly improved. The refrigerated warehouse had become a part of a huge automated system that delivered goods from around the world to stores across a country. "Cold store evils can generally be traced to the design and construction of the stores themselves," explained a British refrigerating engineer in 1948. "With the older stores little can be done about design and construction. The more modern stores are designed to eliminate these evils."[95]

In the United States, newer buildings stretched out horizontally rather than vertically. They also contained enough machinery to almost entirely automate the storage process. In 1949, packages of frozen food at one Dallas warehouse could move through the entire system without being touched by a human hand. An important technology that made that possible was the forklift, which first appeared during World War II and later was widely adopted in private warehouses after the war ended. By cutting labor costs and increasing efficiency, forklifts made the giant warehouses of today possible. Mechanization also vastly increased the amount of both perishable and nonperishable goods throughout the country. "Preservation by refrigeration has become a habit [in the United States], as a result of the much more extensive cold chain linking producer and consumer," reported a group of European executives in 1952 after having visited facilities in America along every link of that chain.[96] The chain has grown even more effective down to the present day.

A small sampling of contemporary facilities offers a glimpse into the sophistication of and wide array of uses for modern cold storage in the United States. Near Springfield, Missouri, Kraft has fitted part of a working limestone mine as a giant cheese cave. The natural temperature of the cave and therefore of the facility stays steady at 58 degrees Fahrenheit; refrigeration equipment lowers it further to 36 degrees. In the Bronx, the firm Master Purveyors runs a 16,000-square-foot meat locker for much of the prime steak consumed in the greater New York area. Thanks to technological advances, the humidity level remains a languid 80 percent in order to prevent the growth of pathogenic bacteria as the steaks age. Banana Distributors of New York operates a storage facility where bananas shipped into New York City are artificially ripened. Since "the energy coming off a box of ripening bananas could heat a small apartment," the facility requires both extensive refrigeration and ventilation.[97]

The size and efficiency of these kinds of facilities make it possible for cold chains to develop on a global scale for many different products. Today, another type of cold storage has also come into its own—namely, refrigerated cases in supermarkets and grocery stores. After all, refrigerated trucks do not just pull up to people's houses to drop off their wares. Perishable foods require refrigerated displays for customers to see and purchase them. Developing the technologies needed to make that possible took time too.

Retail Refrigeration and the Making of the Modern Supermarket

During the late nineteenth century, Americans either bought their groceries from neighborhood stores or from peddlers who traveled from door to door.[98] Neither of these means of purchasing food made much use of refrigeration. Neighborhood groceries seldom had more than a token stock of fresh fruits or vegetables. The few that existed carried just local produce, available only in season. Most of this produce was the sort that stayed fresh without refrigeration, like cabbages or potatoes. Even meat transported to retail butchers faced a continual risk of spoilage because ice could not keep most of the product shipped from Chicago free from bacteria. Reliable electric refrigerated cases did not appear on the American scene until the mid-1920s.[99] Only then did the retail grocery business begin to evolve into an effective distribution system for the products that the cold chain could effectively preserve up to their point of sale.

Supermarkets, large chain stores that sold a greater variety of food than local grocers, arrived in most cities during the Great Depression, offering the cost savings that most consumers needed during those years. Starting with King Cullen on New York's Long Island during the early 1930s, supermarkets grew during the postwar era both in number and in the number of items they sold.[100] In 1932, approximately 70 percent of the food sold in New York City got distributed and sold the same way it had been during the previous century—through a mix of wholesalers, brokers, public markets, and small stores.[101] Modern supermarkets came about because of the efficiencies of mass buying and mass distribution. These efficiencies made it possible to distribute perishable food on a greater scale, and less time in transit meant less opportunity for spoilage. During the 1940s, supermarkets provided a crucial segment of the infrastructure needed to distribute frozen foods.

Only after World War II did the refrigerated cases in supermarkets become anywhere near as effective at preserving food as the rest of the cold chain. The earliest grocer's display cases worked like iceboxes did, most having small doors at the top where clerks or customers could reach in for needed items. While ice could keep perishable food from spoiling, these units had the same drawbacks that using ice in warehouses did. Starting in the 1920s, grocers began to convert their icebox-style cases to mechanical refrigerators using the same technology that homeowners did to modernize their iceboxes.[102] They did not work well. Meat storage cases, for example, usually stayed only a few degrees cooler than the room temperature, resulting in bacterial growth.[103] As late as 1960, many supermarkets still used open display cases. Even when such cases had effective refrigeration systems, the motor had to be run longer and harder to keep the products at the top of such cases fresh.[104] A 1963 textbook on running supermarkets noted that "the last word in modernization" for these establishments included "meat cases, dairy cases, and in most stores, produce cases are fully refrigerated."[105] Even then, the cold chain remained incomplete. Many perishable food products needed special cases in order to be both preserved and seen. Creating and installing those cases around the entire supermarket took both time and money.

Because of these factors, the sale of frozen food did not become profitable until decades after its development during the 1920s and 1930s. When the first retail freezer hit the market in 1928, it cost as much as the rest of the typical grocery store. But by the early 1930s a newer freezer developed by the American Radiator Corporation cost only $300. General Foods rented out these freezers to grocery stores in towns where they test-marketed their frozen food.[106] Despite these improvements, a lack of display space remained the most serious impediment to the growth of the frozen food industry as late as 1950.[107] Open display cases, common in supermarkets as late as 1960, made it easy for consumers to pick out boxes but tended to allow partial thawing of the boxes on top, which were more exposed to the open air.[108] The grocery industry, which depended on perishable or frozen food for 54 percent of its sales dollars in 1954, needed infrastructure improvements in order to prosper.[109] Those improvements only came later in the twentieth century.

Household refrigeration advanced much further and much faster than retail refrigeration did during the early twentieth century. In the early days of the cold chain, home refrigeration came only in the form an icebox. People accepted a primitive technology in their homes because the cold chain that transported so much food so effectively would have been worthless without a way preserve all that food at home until it could be eaten and before it spoiled. An icebox (and its successor, the electric household refrigerator) served as cold storage for the home. Iceboxes form the final link of the cold chain on the route from production to consumption and were the last link in that chain to develop into their modern forms. Nonetheless, their prominence in the household makes them historically significant for both technological and cultural reasons.

CHAPTER 6

"Who Ever Heard of an American without an Icebox?"

In 1875, the Blaisdell & Burley Company of Sanbornton, New Hampshire, introduced its "Patent Elevating Refrigerator"—a cold storage cabinet we would now call an icebox, since it was not electric.[1] Unlike other iceboxes of that time, this one had a gimmick. When the owner turned a key, a system of weights and pulleys allowed it to spring up from the basement through the floor! A flyer for the product explained its intended effect this way: "The family and their visitors are seated at the table when Mrs. A. discovers that an important article has been omitted. Turning in her chair, she touches a spring and the Refrigerator appears, from which the required food is supplied. The surprise of the visitors is easier imagined than described."[2] Most iceboxes had no such gimmick. They sat in kitchens or cellars, keeping food cold as long as they stayed closed. Since any icebox could preserve food to some degree, this company depended upon its novel location in order to promote sales.

The Patent Elevating Refrigerator serves an extreme example of a tendency that underlay nearly all icebox marketing. These appliances did not compete on how well they kept food. Circumstances could vary so widely that even a good icebox might prove ineffective when used badly or if operated in a particularly warm climate. Iceboxes generally competed on the basis of features, only a few of which actually related to food. Under similar circumstances, iceboxes kept some foods well and others not well at all. They also had many inherent inconveniences. Still, the ability to keep any perishable food around in the kitchen longer than its ordinary shelf life made iceboxes very valuable to consumers.

You would not know this by reading icebox advertising. The location of the Patent Elevating Refrigerator saved women "thousands of weary steps," according to Blaisdell & Burley. A similar but non-elevating icebox, the Perkins Upright refrigerator, gave you "access to your ice at all times."[3] Any icebox saved women time, since it alleviated the need for her to perform at least some of her traditional duties related to food preservation, but it also created other problems, like the need for laborious cleaning. Despite such problems, iceboxes

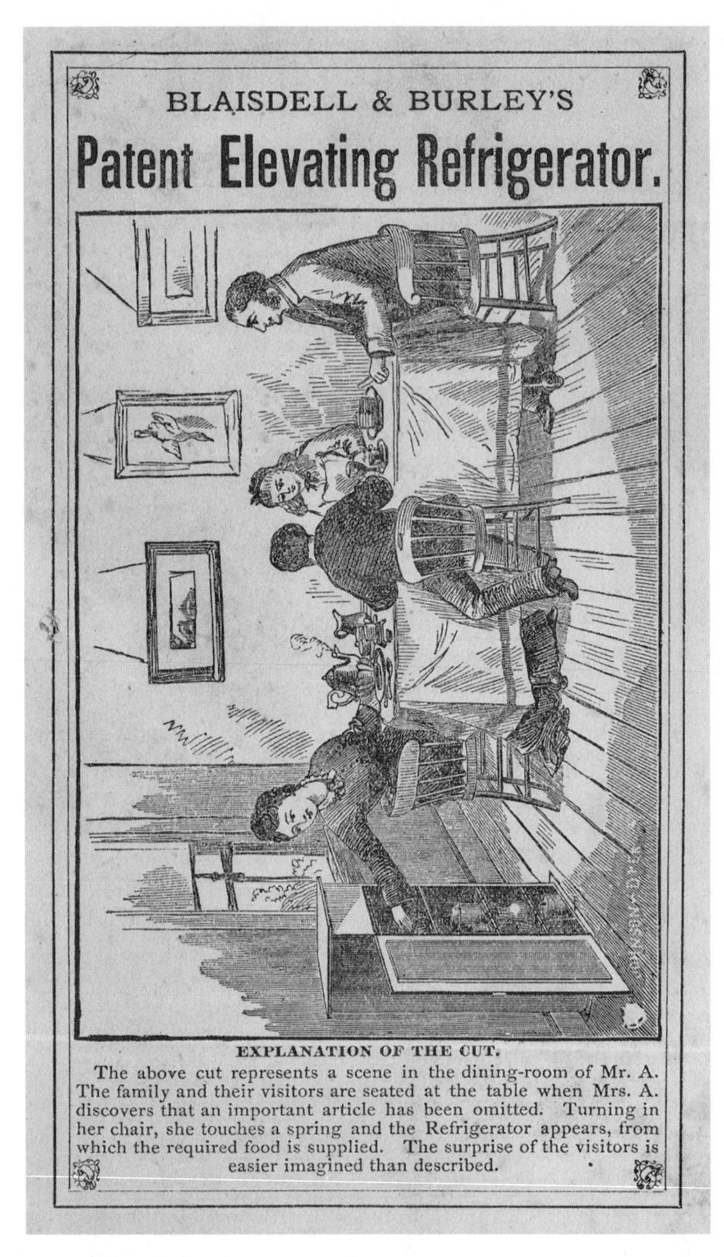

Most iceboxes remained in the kitchen or cellar, but in this ad a woman reaches for her "Patent Elevating Refrigerator" while still at the dining room table because it could rise from the cellar through the floor at the turn of a key. That companies would build and market iceboxes with this expensive kind of distinguishing feature suggests the cutthroat competition in the icebox industry. *Blaisdell & Burley's Patent Elevating Refrigerator*, 1876. Courtesy of Trade Catalog Collection: Baker Library Historical Collections, Harvard Business School.

allowed people to protect some perishable foods like milk and meat without affecting their taste (at least when preserved successfully). Traditional methods could never have preserved these foods in their original form. Iceboxes therefore served as an important transitional stage on the path to the modern cold chain. They did not offer the precision and control of electric household refrigeration, but consumers still preferred to have them rather than having no household refrigeration at all.

Once the cold chain made it possible for consumers to preserve different kinds of perishable food at home, iceboxes became increasingly important for both convenience and the prevention of waste. The convenience came not from their location in the household but from the fact that iceboxes existed at all. In the late nineteenth century in urban industrializing America, people had increasingly less time to shop. People in rural America during the same period did not have the time to travel to markets regularly. Iceboxes also saved time in other ways. Placing vegetables in an icebox took much less time than pickling them. Thanks to iceboxes, preserving leftovers that had not been preserved in other ways became a viable option for the first time. Of course, iceboxes made it possible to keep food much longer without having it spoil before it ever got prepared.

The icebox could have developed only in America. Americans used it because of the uniquely large market for ice that sprang up in the United States during the mid-nineteenth century. By the early 1900s, iceboxes had become part of American culture. "Who ever heard of an American without an icebox?" asked the British travel writer Winifred James in 1914. "It is his country's emblem. It asserts his nationality as conclusively as the Stars and Stripes afloat from his roof-tree, besides being much more useful in keeping his butter cool."[4] But the labor and the cost associated with them might explain the relative scarcity of iceboxes in other parts of the world. In countries that neither produced nor delivered ice for household consumption (in other words, most countries around the globe), iceboxes did not have the same impact. These countries used the food preservations methods that Americans had used before iceboxes arrived.

From Root Cellars and Ice Pits to Iceboxes

Iceboxes seem easy to understand: ice went into a box to keep food in another part of that box cold. While accurate to a degree, that simple description does not convey all the quirks inherent in the use of this technology. Not all iceboxes were the same. Subtle differences in the wood used to make the chest or the type of insulation used to protect the ice had a huge effect on how well an icebox worked. Besides structural problems, housewives could not just put food in an icebox and expect it to remain fresh until whenever they took it out. They had to follow rules; otherwise, even the best icebox would not work properly. Despite the inherent limitations with this technology, iceboxes sold well in the United States throughout the late nineteenth and early twentieth centuries. Because of technological improvements and an increasing variety of choices, the total value of the iceboxes manufactured in the United States increased from $4.5 million in 1889 to $26 million in 1919.[5] Iceboxes allowed the owners to take advantage of all the benefits that the cold chain had to offer at that time. As a result, iceboxes changed people's lives during the nineteenth century, so much so that no family wanted to do without one.

A Maryland farmer named Thomas Moore patented the first ice chest in 1803. He did not enforce the patent so that others could learn from his experience.[6] His pitch for his invention suggests that he intended it to have many of the same uses that refrigeration does now:

> Every housekeeper may have one in his cellar, in which, by daily use of a few pounds of ice, fresh provisions may be preserved, butter hardened, milk, or any other liquid preserved at any desired temperature; small handsome ones may be constructed for table use, in which liquids, or any kind of provisions may be rendered agreeable, as far as it is possible for cooling to have that effect. Butchers, or dealers in fresh provisions may in one of these machines, preserve their unsold meat without salting, with as much certainty as in cold weather; and I have no doubt, but by the use of them, fresh fish may be brought from any part of the Chesapeake Bay, in the hottest weather and delivered in Baltimore market in as good condition as in the winter season.[7]

Although Moore used his invention for transporting butter, it operated on the same principles as a cold storage unit, as his litany of possible uses suggests. The ability to transport products derived from its small size, not its purpose. Over time, iceboxes got bigger and therefore became stationary. Moore's invention might have caught on more quickly had not pickling, drying, and canning met the needs of most early-nineteenth-century eaters. They had no need to try something as ineffective and potentially as expensive as ice refrigeration in the home.

Before iceboxes, American families had root cellars to keep their perishable food fresh. They dug these rooms under or near their houses, below the frost line. Root cellars offered limited protection for grains, potatoes, vegetables, and even apples against spoilage. Unfortunately, they offered no protection against extreme cold or vermin. These drawbacks made the spoilage of at least some of the food stored in a root cellar practically inevitable.[8] Advice books from the early nineteenth century suggested that hard work by women could limit the damage this flawed technology caused.[9] Because iceboxes were smaller than root cellars, limiting the necessary cleaning to a small space must have seemed like an improvement in the early years of icebox technology. Despite this improvement, many inventors eyed the convenience of electric household refrigeration in the late nineteenth century, long before that technology proved viable.

Ice pits predated root cellars; they even existed in ancient times. Archeologists have found ice pits at Jamestown, the first permanent English settlement in America, dating from the seventeenth century.[10] In the eighteenth century, the rich used them the most. Large mansions had more elaborate icehouses attached to them to keep ice that owners cut from local bodies of water. "Opened the well in my Cellar in which I had laid up a store of Ice," wrote George Washington in his diary in 1785, "but there was not the smallest particle remaining.—I then opened the other Repository . . . in which I found a large store."[11] As Washington's experience suggests, early Americans submerged their icehouses in pits, since below-ground temperatures are cooler than surface temperatures. Later, in the nineteenth century, rich homeowners built icehouses entirely above ground because that made it easier to drain off the warmer meltwater.[12]

By the mid-nineteenth century, the rich (and some members of the middle class) began to buy their first stationary iceboxes. Unfortunately, those early iceboxes did a very poor job at keeping anything cold because they had no moisture barrier. This meant that water from the inside of the box gradually rotted the wood and gave rise to horrible odors as it permeated the structure of the appliance.[13] Most of these early iceboxes also had poor insulation, oftentimes nothing but wool or wood shavings between two panels of wood. As 80 to 90 percent of the heat that got into an icebox did so through the walls, poor insulation led directly to poor ice conservation.[14]

Because of the inefficiency of these iceboxes, many inventors patented new designs to improve ice economy. While the U.S. Patent Office's compilation of inventions and designs from 1790 to 1847 lists these innovations in the section on "household furniture," no evidence suggests that these devices made their way into most people's homes before the 1840s.[15] Even then, most Americans built their own icebox if they had one at all. Shortly after that date, a correspondent to the *Cleveland Daily Herald* suggested that rather than build an icehouse, families should instead build "what is called a Refrigerator," which he described as a box within a box with pulverized charcoal between the boxes to serve as insulation.[16] In 1869, the famous household management writer Catharine Beecher provided instructions for building a refrigerator out of a barrel.[17] Iceboxes had become common by that time, but Beecher's suggestion demonstrates that the manufactured iceboxes that resembled pieces of furniture had not yet won the day.

Aficionados generally recognize D. Eddy & Son of Boston as the first icebox manufacturing company. Carpenter Darius Eddy founded the firm after he began making iceboxes for a mass market during the 1840s. Eddy's copyrighted slogan, "The Father of Them All," reflected the firm's pioneering activities.[18] Soon many icebox manufacturing companies joined the industry, but only a few proved successful over the long haul. Many of these manufacturers sold their wares through retail outlets across the country under different brand names. Others only sent their models to their customers through the mail or marketed them through wholesalers like Sears. The growth of one of these firms, the Baldwin Refrigerator Company of Burlington, Vermont, reflects the growing popularity of this appliance. In 1881, the firm had between six and ten men building iceboxes on the top floor of a partner's cheese factory. They made about one hundred that year. By 1886, the firm employed between seventy-five and a hundred men making iceboxes in several factories. Many other icebox manufacturers had diversified businesses. The Ranney Refrigerator Company started its corporate life manufacturing starch from potatoes.[19] This kind of diversification indicates what manufacturers perceived as limited growth potential in the early days of the industry. Few entrepreneurs just built an icebox factory the way they might build a cold storage warehouse after the turn of the century because demand for iceboxes remained low. The craftsman mentality dominated the industry. Sales of iceboxes often depended more upon the elaborate woodwork of the cabinet (carved lion heads, scrolls, etc.) than on their ability to keep food cold.[20]

The most important innovation that these craftsmen introduced involved better air circulation. "I am aware that various modes have been tried and used for circulating air in refrigerators," wrote a correspondent to *Scientific American* in 1855, "but I am not aware, in any instance, [that] a complete and continued rotation, purification, desiccation and refrigeration of the whole of the contained air has been compelled, as it is in my invention."[21] Despite such claims, credit for recognition of the importance of air flow in ice refrigeration has gone to John Schooley, who began experimenting in 1848 with iceboxes that separated the ice and the food in separate compartments. In *Scientific American* in 1856, Schooley claimed that by "combining an ice receptacle with the interior of a refrigerator . . . a continuous circulation of air shall be kept up through the ice in said receptacle, and through the interior of the refrigerator . . . so that the circulation air shall deposit its moisture on the ice every time it passes through it, and be dried and cooled."[22] All subsequent manufactured iceboxes copied this design to some degree, especially the use separate compartments and a concern for airflow.[23] The first iceboxes with ice loaded at the top appeared in 1856 too. This arrangement improved air circulation throughout the entire chest. The cool air dropped and the warm air rose, bringing heat from the food compartment (or compartments) with it.[24] From the 1850s onward, schematic diagrams became increasingly common in icebox promotional literature, each with arrows representing the path of air traveling across the ice and throughout every compartment of the model in question.

The changing location of iceboxes in the home suggests their increasing importance and general improvement over time. Large cabinets for storage rather than for helping with the food preparation process did not fit well in cramped kitchens. In 1856, Catharine Beecher recommended keeping the icebox in the cellar, the same place from which the "Patent Elevating Refrigerator" emerged.[25] Towards the end of the century, this conventional wisdom changed. "Some people think that the cellar is the proper place for the refrigerator because it is cool," wrote C. H. Leonard of the Leonard Cleanable Refrigerator Company around 1890. "This is a great mistake because the air is damp, and you cannot preserve the food in a damp atmosphere. Besides, the refrigerator is apt to swell out of shape in the cellar, just as your piano would," both being made primarily of wood.[26]

The Problems with Iceboxes

Moving the icebox into the kitchen would not have done much to improve the appliance's food preservation capability. Even later-model manufactured iceboxes required that users have a full understanding of their limitations in order to operate them successfully. Any icebox could keep food cold—for a while—but preserving food effectively and economically required addressing many problems. For example, icebox owners had to watch the drip pans under their appliances closely in order to keep them from running over. "I remember my aunt pulling a white enamel drip pan from the icebox, holding it in front of her and moving step by careful step across the kitchen floor to the kitchen sink, the frigid water shifting dangerously in her arms," recalls the writer Alice Demetrius Stock. "I remember watching the water splosh in a great cascade all over her feet and all over the floor and the 'funny words' she shouted out."[27] Since opening the door to check the ice pan sped up melting by allowing heat to get in, homemakers needed to have an intuitive sense of when to change the ice.[28]

After the start of the twentieth century, iceboxes went from having drip pans to being connected by pipes to outside drains. Pipes instead of pans created new problems. The pipe had to be cleaned regularly with a long brush so that food residues would not clog it. "This slimy substance must be washed out," explained the Northern Refrigerator Company in its 1915–16 catalog, "as it only takes a month or so for the waste pipe to be completely filled with it; then the melting ice will find its way into the cold air passage and the provision chamber will be flooded."[29] Every waste pipe also required a trap at the end of it. Otherwise, poisonous sewer gases might rise up through the pipe and taint the food in the icebox.[30] Other icebox designs had the runoff drain into a sink in the basement in order to avoid this same problem.[31] Warm air could also rush in, melting the ice and spoiling the food.[32] Users might have wished to have their iceboxes hermetically sealed, but then they would never have gotten at the food they stored there.

Inside, the cabinet required even more maintenance. Sister Loretto Duff, writing shortly before electric household refrigeration took hold, offered an excellent primer on the difficulties inherent in keeping the inside of an icebox clean. "All waste and overflow pipes of refrigerators . . . become foul with grease, dust, lint, and other substances, if not well cared for, and frequently contaminate the air of the whole house," she explained in 1916. "Put not hot food in the refrigerator, nor food with a strong odor, as salt fish, etc. Examine the refrigerator daily, that no food be left to spoil or cause bad odors. A drop of spilled milk or particle of other food thus left will contaminate a refrigerator in a few days." So like root cellars, iceboxes required regular, thorough scrubbing and cleaning. "The refrigerator pan should be washed every day or two in boiling water," Duff wrote. "Once a week, when the ice supply is low, everything should be taken out and every part of the refrigerator washed with a hot solution of washing soda. . . . Clean the pipe its entire length with a cloth pushed through by means of a skewer or a piece of wire. Do not neglect the space under the ice rack. Wipe all parts as dry as possible. Place shelves near the fire to dry, or, better, in the sunlight or open air." Long-term storage was impossible, since the icebox had to be left open to "thoroughly dry; otherwise it will become moldy."[33]

This was particularly true of wooden iceboxes, common in the early days of this technology and still sold later as cheaper models. They tended to grow rank over time as melted ice mixed with food particles permeated their inside surfaces. No amount of cleaning could remove that smell. The Leonard Company manufactured cleanable refrigerators with removable linings to make cleaning easier, the first patent along these lines coming in 1882. The Monroe Porcelain Refrigerator advertised: "All corners rounded; no joints or crevices where food may lodge and decay. Light in every corner; as easily cleaned as china."[34] The Bohn Syphon refrigerator claimed that it "practically cleans itself."[35] As these easy cleaning features existed only on higher-end icebox models, most housewives had to keep on scrubbing.

Even with thorough, regular cleaning, hygiene remained a problem for even the most expensive iceboxes. The cleanest iceboxes spread germs and disease. The Herrick Refrigerator Company of Waterloo, Iowa, tried to convince people to buy its expensive zinc-lined iceboxes, arguing that "all the food, the very substance upon which you exist, passes through the refrigerator." "Do you not," its catalog continued, "consider health in preference to money or any thing else? . . . Your health is more precious than money. Parents should respect the hygienic truth for the baby's sake, even though they do not for themselves."[36] But foreign matter in the melting ice could foul an icebox whether it permeated the lining or not. "Melting ice absorbs impurities," reported the *New York Tribune* in 1915. "If the water in the drip pan is analyzed by a competent chemist he can name all the articles that are stored there."[37] Catherine Owen, in her 1889 manual *Progressive Housekeeping*, recommended keeping fish, bacon, ham, cheese, or any kind of cooked vegetable outside the icebox because they could taint the appliance with a permanent smell.[38] Natural ice proved particularly difficult to use in iceboxes, since the sediment in it tended to clog the pan where the water collected or the drain on its way out. That explains why the Boston Scientific Refrigerator Company warned its buyers to "always wash your ice before placing it in the refrigerator," thereby giving housewives yet one more thing to do.[39]

As was the case with foods in cold storage, foods inside an icebox had the potential to taint other foods with strong smells. "It is not possible to keep a cantaloupe in the same compartment as the butter, even though it be in a covered stone crock," suggested the *Philadelphia Inquirer* in 1893, "without the butter becoming tainted. Nor can poultry, raw or highly spiced meat, be placed near unsealed milk without distinct detriment to the latter."[40] Of course, there were no plastic storage containers at this time to solve such problems. The possibility of cross-contamination explains why the *Idaho Daily Statesman* noted that "up to date" residences had separate refrigerators for milk and butter and meat.[41] Since most families had enough trouble purchasing and icing one icebox, they probably did not enjoy their food as much as people who could afford to follow this advice; a single icebox limited the families that used it to refrigerating only those foods in the most dire need of preservation.

Aside from the various drawbacks of icebox technology and maintenance, the cost of purchasing ice on a daily basis could make operating iceboxes very expensive. Some households required as much as a hundred pounds per week. This inability of early iceboxes to preserve ice goes a long way towards explaining why U.S. ice consumption increased rapidly from the mid-nineteenth century forward.[42] The best iceboxes had to make the ice melt slowly to conserve costs—but not too slowly; otherwise, it would not keep anything cold. Ice could keep food cold only when it was melting, since this is what chilled the air inside the icebox. Homemakers concerned with the cost of ice tended to skimp on it. Sometimes, a family might skip buying ice entirely for the six cooler months of the year, counting on the temperature down in the cellar being cool enough to preserve food without ice.[43] Because of such shortcuts, one household refrigerator manufacturer explained, "the butter is only tainted. The meat requires an additional pinch of soda to 'make it sweet' again. The melons and peaches are soft in spots. The berries need careful picking before they are served." Yet the frugal housewife "sees no relation between her ice economy and the amount of spoiled food she has to throw away."[44] Such a housewife undoubtedly blamed her icebox rather than herself for this situation.

Many housewives wrapped their ice in paper or a store-bought ice blanket (usually just a waterproof cloth) to make it last longer. Perhaps this was a holdover from the early days of iceboxes, but wrapped ice prevented the temperature inside an icebox from getting as low as it would have if the ice had remained exposed. This meant that families kept ice longer, but it also meant more spoiled food. "When ice doesn't melt it isn't doing its work," explained the *Tribune* again. "Insulate it completely so that it doesn't melt at all, and you might as well fill your ice chamber with Belgian paving blocks."[45] Insulation in the walls of the refrigerator kept the cold in to prevent the ice inside the compartments from melting. Unfortunately, the poor insulation in most cheap manufactured refrigerators did not do this job well at all: it did not create temperatures low enough to preserve most foods well because condensation built up inside the compartments. This provided a wonderful environment for festering bacteria.[46] Preservation by icebox, in short, did not always work as well as icebox owners wanted.

The use of ice to preserve food in the home also presented logistical problems. Miss delivery one day or even get a late delivery, and all your perishable food might spoil. Even if you let the ice melt too far, you would have to keep the door closed for a long time after re-icing to get the temperature of the food compartment down to a level that could preserve food. And, as mentioned before, open the door too often, and the refrigerator would not stay cold.[47] That might explain why those families that did own early iceboxes tended not to store food for them for long, preferring to use them just to keep fresh foods from going bad.[48] Home refrigeration would only become an important part of daily life when families could store their food effectively for long periods of time.

Ice also took up considerable space in the cabinet, thereby limiting the amount of food that could fit inside. Writing for *Harper's* in 1866, Catharine Beecher argued, "The ice closet is better than a refrigerator, as it can be used for larger quantities of meat and milk, and for many other things in hot weather."[49] If a housewife put too much food in an icebox, the air would not circulate properly. Food in iceboxes also released heat. Therefore, too much food made the temperature go up, endangering the freshness of everything inside. To prevent this from happening, people had to buy bigger iceboxes, thereby foreshadowing the huge American household refrigerators of today.[50] Without a bigger icebox, things which people might have liked cold either did not stay cold or were not purchased in the first place.

To address all the criticism their products received for many problems in operation, icebox manufacturers offered elaborate instructions for an appliance that seemed so simple. The instructions accompanying D. Eddy & Son's icebox from 1894 deserve attention, since they cover all potential problems:

> New refrigerators should stand from twelve to twenty-four hours after the ice box is filled before replacing articles of food in the provisioning chamber.
>
> The ice should be washed and put in carefully. It should not be thrown in, and *it should never be wrapped in anything*, as that prevents the circulation of cold air. In

> getting ice from the refrigerator for table use, be careful not to pick off more than is needed, as the ice will melt more rapidly in small pieces.
>
> The ice apartment of a new Refrigerator, or of an old one at the beginning of a new season, should be entirely filled with ice, in order that it may become thoroughly cooled. Never let the ice get wholly out before replenishing.
>
> The strainer should be constantly kept over the water outlet inside the Refrigerator, so as to prevent the escape of cold air through the waste pipe.
>
> The covers and doors must be kept shut. They should never be left open, and never slightly ajar, as is very often the case.
>
> The zinc lining should occasionally be washed with soap, and warm water, and then wiped perfectly dry. This will keep the Refrigerator clean and free from odor.
>
> The food should never be placed in the Refrigerator in a warm state; for anything warm will cause dampness and moisture.[51]

Such a list demonstrates the limitations of this technology. Yet, if we leave aside the instructions dealing with the use of ice, such rules could also apply to modern household refrigerators.[52] We have yet to discover a foolproof method of food preservation, but consumers still accept occasional spoilage and the need for refrigerator maintenance because of the many benefits that refrigeration brings.

"Americans are without doubt a most extravagant people in their mode of living," wrote the refrigerating engineer Justus Goosman in 1924. The icebox "is often overstocked, and since its temperature is not low enough to conserve food beyond a certain period, much of it spoils and is finally carried off by the garbage wagon."[53] Despite such difficulties, the convenience of even far-from-perfect refrigeration eventually made iceboxes of some kind a household necessity, particularly when ice became cheap enough so that nearly every American could afford it. Iceboxes had very significant effects on those families that bought them, allowing people to switch from a diet based on salted meat and bread to one that included fresh fruits and vegetables.[54] Families that did not own an icebox but still wanted ice left the excess each day in sawdust or straw to insulate it until it melted.[55] They could also follow the advice offered by the *Farmers' Cabinet* of Amherst, New Hampshire, which in 1865 had recommended just wrapping "ice in a flannel sheet, or any large piece of woolen cloth or dry blanket" to slow its melting. According to the author of that advice, ice lasted for two days in summer using this method.[56] People's willingness to go to such lengths for ice that would not melt quickly demonstrates an enthusiasm for ice and iceboxes in America that did not exist elsewhere around the world.

Iceboxes around the World

Although by far most ice in England went towards industrial uses, the owners of large manor houses often had their own icehouses built, using a model with a circular building over a chalk, gravel, or sand pit.[57] The distributors of Wenham Lake Ice, which shipped ice to London first from Massachusetts and then from Norway, did sell refrigerators. The company described them in 1867 as "exceedingly well-adapted for all domestic purposes." The same advertisement noted that ice in such boxes could last up to a few months.[58] If true, then the firm's London customers certainly did not open their iceboxes anywhere near as often as Americans did. Most likely, too few customers bought iceboxes to put that assertion to the test.

After Wenham Lake Ice disappeared from London around 1890, the cold chain in Great Britain significantly degraded. By the end of the century, the limited delivery service that had been available disappeared.[59] All the imported ice came from Norway through east coast fishing villages, since fishermen used much more than English consumers did. In fact, few English consumers used any ice at all. An American observer wrote in 1907 that "a well-to-do family in an English town of 100,000 people does not know that there is such a thing as ice, so far as household purposes are concerned. The butcher in the market has a deep pit filled with ice in which he hangs his meats, but he will decline to sell any of [the ice], because it is so much trouble to get at it. Ginger beer and bottled soda are delivered warm and drunk warm."[60] The pubs and taverns that had ice advertised the availability of "American iced drinks," which shows that tourists from the United States constituted the primary market for such beverages.[61]

Since British consumers never took to ice the same way Americans did, they had little interest in iceboxes. In London, the American consul there wrote, iceboxes were used "to a very limited extent for domestic and household purposes . . . because of the climatic conditions and the general comparatively uniform and moderate temperatures." The consul in Leeds offered a more telling observation about iceboxes in this country: "Every house occupied by persons whose circumstances would warrant the outlay necessary to obtain a refrigerator is provided with cellarage, where meat, vegetables and so forth, are kept before being used. These cellars, while below the exterior level of the ground, are generally well ventilated, and, being furnished with stone tables, they serve the purpose of a cold-air chamber, with the great advantage of abundance of space and convenience of storage."[62] With respect to refrigeration, England in 1891 resembled America in 1841.

France had even fewer iceboxes than England did. "The mass of the people [in France] use no ice whatever," reported *Ice and Refrigeration* in 1901. Instead, most French people still used cellars for cooling purposes.[63] With no knowledge of refrigeration technology, the inhabitants found cold storage suspicious initially, but the French attitude toward refrigeration had important cultural dimensions to it. "My dear, those frightful *glacieres* of yours are practically unknown with us," a French housekeeper explained to a contributor for a 1917 issue of *American Cookery*. "It may be that one would find them in use at the big hotels of our country that are going up so rapidly just for you Americans when you come to see us, but in a French family I do not know of even one."[64] This dismissive attitude epitomizes the legendary French haughtiness. This story, like the term "American iced drinks" in British pubs, also suggests the role that Americans played in spreading refrigeration around the world, as they wanted to eat the same way abroad that they did at home.

Unlike any other country in Europe, Germany closely resembled America in the use of iceboxes. Nonetheless, back in 1890, an American consul in the country reported that "except in wealthy families and large hotels and markets, refrigerators are generally of the most primitive form. Even in large hotels I have found ice-boxes consisting simply of a wooden box lined with zinc." But unlike other European countries, Germany had hints of an infrastructure capable of providing ice for domestic food preservation uses. For example, Dusseldorf had above-ground icehouses, which were known as "Americans" (this name strongly suggests whose lead the Germans followed when it came to cold chain infrastructure).[65] As in England, most of the artificial ice went for industrial purposes like brewing beer.

Outside of countries with existing ice industries, even fewer iceboxes existed. "It is no exaggeration to state," reported the U.S. consul in Genoa, Italy, "that more refrigerators can be found in any city of 10,000 inhabitants in the United States than in this entire consular district of over 900,000 souls." He reported that many Italians "believe ice injures rather than soothes the palate and the stomach." Cost formed another obvious barrier—not just that of the icebox itself, but of the continual supply of ice needed in order to make it operate.

Cubans had once bought lots of Frederic Tudor's ice, but not at the end of the nineteenth century. "The fact that on average only three tons of ice per day are consumed in the city of Santiago de Cuba, boasting of a population of nearly 48,000 souls," reported the consul there, "is conclusive proof that the resources of the people do not permit a large consumption of ice. . . . Almost all the ice consumed here is used by Spaniards and foreigners." Around Bombay, the people who consumed ice during the late nineteenth century stored it in simple wooden boxes, recognizing, perhaps, that they got little added value from an expensive icebox.[66] In America, however, the quality of most iceboxes on the market eventually took a turn for the better.

An Icebox Hierarchy

In 1877, the Boston Scientific Refrigerator Company introduced the "Hard Times Refrigerator," a wooden chest big enough to store fifty pounds of ice, for people facing difficult economic circumstances.[67] That Americans facing hard times still wanted refrigeration—even poor refrigeration—speaks volume about the importance of refrigeration in American life. In 1893, the report from the Committee on Awards at the 1893 World's Fair noted the improvements in iceboxes: "The so-called ice chests of twenty years ago have all been relegated to the background, and few households are found willing to purchase the best samples of the refrigerator of ten years ago. Since then the circulation has been made more perfect, and the methods adopted to secure insulation, as well as the many devices used to strengthen the boxes, the locks, the traps, and the material used have combined to make the best refrigerators exhibited at the Columbian Exhibition almost ideally perfect."[68] Despite these innovations, much room for improvement existed. A study presented at the 1913 International Congress of Refrigeration in Chicago found that 177 out of 300 iceboxes examined could not keep food at a temperature lower than 50 degrees Fahrenheit, making them practically worthless for preserving food.[69]

The existence of an icebox hierarchy explains this apparent anomaly well. "In no other single article of household convenience and utility is there such a range of quality and price as in refrigerators," observed the *New York Tribune* in 1914.[70] While the iceboxes in the homes of working people could cost from $10 to $20, those of the well-to-do could cost anywhere from $25 to $150.[71] Over time, the lower-end iceboxes gradually improved. "The old type of zinc-lined refrigerator is almost extinct," reported the *Tribune* in 1919. "They are heirlooms—not to

The Excelsior Refrigerator, marketed by C. H. Roloson & Company of Baltimore in 1877, was intended for restaurants and other public institutions that needed to keep their perishable food fresh. Courtesy of the Hagley Museum and Library.

be purchased by anyone taking a new start in life. This is because they are not easily cleaned and do not wear well, owing to the seams and the corroding and the corroding of the metal itself from moisture." Now, the paper explained, "the choice lies between a box lined with enameled wood, a porcelain enamel on a metal base, a one-piece porcelain lining or a solid porcelain box, named in order of their expensiveness."[72] Wealthy people invested in newer high-end iceboxes.

At least the expensive models had many marketable improvements that justi-

Both the size of the McCray Patent Refrigerator and its finish demonstrate its position at the top of the hierarchy of available iceboxes. JWT Domestic Advertising Collection, Box 1, Ad Portfolio—Database #J0015, Emergence of Advertising Project, John W. Hartman Center for Sales, Advertising & Marketing History. Courtesy of Duke University David M. Rubenstein Rare Book & Manuscript Library, http://library.duke.edu/digitalcollections/eaa.

fied the added cost. Manufacturers usually built them out of porcelain as one-piece units, meaning that these iceboxes did not have seams like older zinc-lined models. Porcelain was resistant to absorbing smells and was easier to clean. Thus, hygiene became a key selling point for porcelain iceboxes. Some firms built iceboxes directly into buildings or private homes so that they could be tailored to the customer's specific needs. Invariably, the customer had to contact the company to find out how much that would cost. A pamphlet from the Monroe Refrigerator Company of Monroe, Michigan, implied that if the buyer had to worry about how much a built-in icebox cost, he probably could not afford one for his home by including a list of names of people who had purchased the company's services on the back. The first three were J. P. Morgan, John D. Rockefeller, and Jacob Astor.[73]

Ironically, the arguments that manufacturers made for customers' spending lots of money on these posh iceboxes prefigured the sales pitches that electric household refrigerator manufacturers made after 1920. "In selecting a refrigerator," argued the makers of White Mountain Refrigerators (which were high-end iceboxes), "its mission to guard the health of the family and the purity and pleasure of the dining table, a serious duty is before the housewife. . . . She knows that true refrigeration, like the pearl, is found below the surface." Electric household models stayed much cleaner since they required no ice at all and therefore produced no mess. The same observation applied to another common feature of expensive iceboxes: rear doors that allowed ice to be loaded from the outside of the home so that "the house can be locked up without waiting for the ice man, and [so] that it is no longer necessary to tolerate tracks or muddy boots and trails of dripping water or sawdust brushed from the ice."[74] This kind of rhetoric suggests considerable gendered tension between burly icemen carrying dirty blocks of ice through clean kitchens and the women who kept those kitchens clean that probably went back to the early days of the industry. The rapid elimination of the iceman soon made those tensions obsolete.

Despite the iceman's eventual demise, getting rid of him looked like a very tall order in the early days of the twentieth century because by that time iceboxes (albeit often very cheap ones) had become extremely widespread. Icebox sales took off during the 1870s, and they became increasingly common in the houses of the lower classes.[75] Even if poorer families could not afford an icebox, they still took ice. Robert Coit Chapin, reviewing the standard of living of working-class families in New York City in 1909, found iceboxes "in more than four-fifths of them," even though "in some cases it is reported that the ice is kept in a tub," reflecting the persistence of that earlier practice.[76] This ice offered relief from the heat. The decision to purchase ice despite no good means to preserve it demonstrates the value of controlling the cold to even the poorest Americans. More Americans had access to some kind of refrigeration than did people in any other country, even at this early time. Keeping ice in the bathtub did not offer particularly good control over the cold, but the fact that even the poorest families preferred having access to ice, even without an icebox, suggests how much they valued it.

The quality of higher-end iceboxes improved greatly once competition from the electric household refrigerator began to emerge.[77] Icebox firms also benefited from the fact that for most of the 1920s, electric refrigerator manufacturers retrofitted existing iceboxes with machinery to turn them into modern household refrigerators. Many electric household refrigerator companies even bought their cabinets from icebox manufacturers. In 1929, however, with the coming of the single-piece, pressed steel cabinet, this relationship ended.[78] Most icebox manufacturers went out of business during the 1930s, casualties of the Depression and aggressive marketing by a technologically superior substitute (even though the iceboxes they built stayed in operation, especially in rural homes, for years afterwards).[79] The last icebox manufacturer in the United States went out of business in 1953.[80] By then, a new era in home refrigeration had already begun.

The Prehistory of the Household Refrigerator

Pinpointing the first electric household refrigerator leads to the same difficulties encountered when asking who invented the modern automobile. The answer to both those questions depends upon how those terms are defined. In truth, no first electric household refrigerator ever existed. Instead, the industry went through a long transition between iceboxes and modern household refrigerators during which these two technologies had many traits in common. In 1929, as many iceboxes remained in use as electric household refrigerators, but as household refrigeration technology improved, the old technology gradually disappeared.

Inventors had dreamed of electric household refrigeration long before the first viable models arrived. "We know of no cheap way of making a refrigerator for household purposes without ice," wrote the editors of *Scientific American* in 1885. "Such a process, if convenient, would be worth a great deal of money."[81] Despite this profit incentive, creating household models faced tremendous technical obstacles. The large electric refrigeration units that made ice weighed at least five tons, but household units had to be small and light in order to fit into most kitchens. Attendants monitored larger machines around the clock, but a household unit could not require constant attention. A successful household refrigerator also needed an inexpensive power source. Besides these traits, it had to be safe and affordable too.[82] Nobody could simply shrink existing mechanical refrigeration systems from the late nineteenth century because of efficiency concerns. Smaller machines generated more friction, which created more heat, which in turn made it harder to create cold.[83] Successful electric household refrigeration makers eventually resolved this problem by using refrigerants other than ammonia that could absorb heat under lower pressure.

All these hurdles did not prevent a slew of inventors from trying to perfect household refrigeration technology. The first attempt to build and market an electric household refrigerator came in 1887. Several models came on the market in 1895, but they did not survive long enough to enter the historical record in any detail. Several other models became available in 1910, but these machines failed on the market too.[84] "The small refrigerator plant, or ice machine, is coming," wrote Mary Pattison in her 1915 treatise *Principles of Domestic Engineering*. "We already have a number from which to choose, but up to date we can not fit them in practically to the average home with assurance of economy; although 'The Montclair' and the 'Automatic Household Refrigerator' seem to be the answer to this need. Both are worthy of investigation."[85] These models met the same fate as their predecessors. The first household refrigerators originated from within the industrial refrigeration equipment manufacturers of the previous century, but these models did not form the basis of a new industry. They were technological dead ends.

The first electric household refrigerator to survive its infancy in any form was called the DOMELRE (an acronym for "domestic electric refrigerator"). Not a self-contained appliance in the modern sense, it was a small device that could be "installed in any icebox." Its inventor, Fred W. Wolf, Jr., the son of an earlier refrigeration pioneer, brought it out in 1914 as an extension of his work building smaller refrigerating machinery. The device had enormous potential. After all, it promised refrigeration without the cost of ice, without the need to clean the icebox, and with low operating costs, since electricity was cheap at this time. Despite these advantages, the company sold only 525 of these units.[86] While the DOMELRE offered more precision in controlling the cold, household refrigeration had not reached the required reliability or affordability threshold for household refrigerators to find a mass market.

In 1916, Henry Joy of the Packard Motor Company bought the rights to the DOMELRE and began to manufacture a new device with the same technology in Detroit. Joy called his new company, and the device, Isko. An early pamphlet described his product: "This wonder machine fits any ice box. It is easily installed, either on top of the ice box itself or in any convenient place and attached to the electric and water systems." As the pamphlet later indicated, "any convenient place" meant that the refrigeration device was installed in the basement and connected to the icebox through a hole in the floor.[87] The Isko machine was only slightly more successful than the DOMELRE, with 1,500 models sold over the life of the company.[88] Isko machines had problems with their motors, and oil bubbles and iron sulfide would eventually clog the pipes. Luckily for Joy, General Motors bought the company for its patents in the vain hope that it could keep competitors from developing a machine that was even somewhat similar.[89] Some companies continued to retrofit old iceboxes with mechanical refrigerating units well into the 1920s. This strategy required many different sizes and shapes of equipment in order to fit into different-sized iceboxes. This made it substantially more difficult to standardize parts and therefore cut costs.[90]

In many ways, the refrigeration industry of the late 1910s and early 1920s resembled the refrigeration industry of the 1880s and 1890s. During the transition from natural ice to mechanical refrigeration, everybody recognized the advantages that reliable, inexpensive mechanical refrigeration offered. Even so, making this transition still took a long time. The same dynamic existed with home refrigeration. The obvious advantages of household refrigeration led many people to try to create household refrigerators. Since it was impossible to shrink the existing commercial technology, developing these new home appliances required new technological breakthroughs. Once household refrigerators became both effective and affordable, they created a revolution in the day-to-day lives and diets of every American who gained access to that technology.

CHAPTER 7

The Early Days of Electric Household Refrigeration

"How do you do, Mrs. Prospect?" So began a 1923 "demonstration" book for Frigidaire salesmen designed to help them sell electric household refrigerators door-to-door. Assuming the salesman gained admittance to the home, his script told him to immediately approach the family's existing icebox with a thermometer. "Mrs. Prospect," went the pitch, "we find that the average ice box maintains a temperature of about 55 degrees, and I think you will agree with me that this will keep food properly for only a short time." After measuring the temperature inside the prospect's icebox and showing her the reading, the salesman then said, "The temperature in your refrigerator is ——— degrees. This is slightly warmer than I expected. If you had *Frigidaire*, the temperature would certainly be ——— degrees colder than you now have in your icebox. . . . Won't you please talk this matter over with your husband tonight as, in all probability, I or one of our men will call upon him tomorrow afternoon and tell him the benefits of owning a *Frigidaire*."[1] To earn the right to give this spiel, a Frigidaire salesman had to memorize it; then he had to learn how to take the current products apart and put them back together again. Then he would go out on the road for two trips, of three weeks each, with an experienced salesman.[2]

Salesmen needed this kind of knowledge of the refrigerator and its differences with iceboxes in order to promote the sale of a product that still had a reputation for unreliability when these instructions were written in 1923. Despite the obvious advantages of electric household refrigerators, their high cost at that time made them a major investment for even the wealthiest of consumers. Worse still, refrigerator owners went without ice if the appliance failed for any reason. Another refrigerator company, Kelvinator, tried to assuage fears about the reliability of its products by linking its household refrigerators to the history of mechanical refrigeration. As a 1922 pamphlet explained, "Every cold storage plant, every packing house, every artificial ice company, and most of the large butcher-shops, hotels, restaurants and hospitals, have been using mechanical refrigeration for anywhere from twenty to eighty years." It argued that "the basic idea behind Kelvinator is not new. It is no experiment. It is no novel idea, which may or may not be practical. It is a time-tried, and time-tested principle in which hundreds of millions of dollars are invested right now and without which our complicated social structure could not exist."[3] By reminding electric household refrigeration customers of the technology already operating in the cold chain, Kelvinator hoped to induce consumers to bring the most modern example of that chain into their kitchens.

To shrink refrigerating equipment down to a size that could fit in a crowded kitchen, engineers (many of whom had no previous experience with refrigeration) had to reconceive the entire apparatus in order to make household refrigeration both reliable and safe. Unwilling or unable to think so far outside of accepted technological norms, the earlier manufacturers of large-scale, multi-ton refrigeration units never entered the household market despite the obvious potential for such machines. Since they only custom-built large machines for individual customers, they could not shrink the existing technology to fit the home kitchen, and they proved unwilling to retool their production processes to meet the substantial demands of mass production. Besides changing the production process, entering the home refrigeration market would have required a distribution and marketing system to sell appliances to the consumer market rather than just sell to other firms.

Some reconceptions of refrigeration units to household size worked better than others. Between 1924 and 1934, the physicists Leo Szilard and the even more famous Albert Einstein applied for twenty-nine German patents, most of which dealt with home refrigeration. Electromagnetism powered their system by drawing a metallicized refrigerant around a system that had no moving parts. Because of noise, their system never left the prototype stage.[4] Other new entrants into the refrigeration industry with new ways of thinking had to go through another long period of trial and error before this new appliance became commercially viable. Along the way, every manufacturer of household refrigerators had many development costs. Despite their reliance on the same refrigeration cycle that had made industrial-size machines possible, household refrigerator manufacturers essentially started a brand new industry.

Before the introduction of home refrigeration, few people outside the industry had ever witnessed mechanical refrigeration at work because only industrial consumers used this technology. As late as 1924, the journal *Electrical World* reported that "people as yet are skeptical as to the feasibility of electric refrigeration in the home."[5] The *New Yorker*, perhaps the most urbane magazine of any era, marveled in the middle of the decade about the new machine's ice-making capability: "A little water is put in some mysterious place [inside the refrigerator]; a few minutes pass, a magic door opens, and a tray of small ice cubes appears before startled eyes."[6] Early electric household refrigerator advertising demonstrates this attitude well. Many ad illustrations depict small groups of people clustered around open refrigerator doors. This iconography suggests the Magi and the Christ child, especially because of the amazed expressions on the faces of the characters. Their amazement came not from the food inside the refrigerator but from the fact that the refrigerator worked at all.[7] This part of the cold chain's infrastructure briefly served double duty as a household novelty until people learned to take home refrigeration for granted, in the same way that they also took the rest of the cold chain for granted.

The early sense of wonder about electric refrigeration suggests that the household refrigerator industry had to sell the idea of its technology before it could sell the product. Early in the 1920s, consumers had good reason to question the reliability of household refrigerators. The appliances cost a lot of money, preserved food poorly, and often required multiple service calls in order to operate properly. This explains why so many early refrigerator companies either went bankrupt or got bought out by larger firms. As long as people were astonished that electric household refrigeration could actually work, they could never trust it completely. Home refrigeration faded into the background only when it became as reliable as the equipment in an ice plant. Once the machines became more reliable and less expensive, the home refrigerator market expanded quickly. Only the largest, most experienced companies could create machines that ran well and cost less than earlier models. These firms championed the technological breakthroughs that came to define the modern refrigerator of today. They also made it possible for all the food that the cold chain could transport to be stored safely inside the home for the first time.

"It Is Easier to Lose Than Make Money"

For more than five months in 1923, a General Electric engineer named Alexander Stevenson, Jr., compiled a lengthy report for his company's president, Gerard Swope, on the possibility of their company's entering the domestic refrigerating market. He recommended that the firm proceed with caution. "No thoroughly reliable and reasonably inexpensive domestic refrigerating machine has yet to be developed by any of the major companies engaged in this enterprise," Stevenson wrote. "During the developmental stage, it is easier to lose than make money. Therefore, the General Electric Company should not undertake this business in the hope of immediate profits." The same year that Stevenson wrote his memo, the National Association of Ice Industries claimed that refrigerator companies had already spent $16 million trying to develop these appliances. Stevenson disputed that figure, but investors and entrepreneurs certainly lost millions trying to establish this technology because of the large number of obstacles that they had to overcome for it to take hold.[8]

The successful electric household refrigerator had to be small and light enough to fit inside a typical dwelling and reliable enough to run automatically (since, unlike the earlier multi-ton industrial machines, there would be no refrigerating engineer to watch over it). It had to have a power source that did not introduce a lot of additional complications.[9] Many early refrigerators used noxious or even poisonous gases as refrigerants; they therefore needed to be leak-proof to ensure consumers' safety. These appliances could not cost too much if they were ever to displace iceboxes in American kitchens.[10] Consumers also wanted their machines to run quietly. "The importance of quiet operation cannot be overestimated," wrote Stevenson. "People do not especially object to a noisy vacuum cleaner or a noisy washing machine because these are run during the day. But, most of these refrigerating machines start up at almost any time and the slightest noise in the dead of night is very disagreeable. People may get used to this, but there will be a sales resistance until they do."[11] While appearance was hardly as important as some of these other concerns, the failure of the first stand-alone refrigerators to conform to the aesthetic design standards of the best-looking iceboxes (let alone the sleek, multicolored refrigerators of later decades) offered housewives one more reason not to switch to this still untested technology.

The failure of manufacturers to standardize their refrigerators caused yet more problems. This made them very hard to fix during frequent service calls. The costs associated with these calls practically destroyed the electric household refrigerator industry during its early years. Kelvinators required two service calls per machine on average as of 1923. That same year, typical Frigidaires needed five calls for repair during the first year after purchase.[12] Dissecting the books of one early failed household refrigerator manufacturer, *Ice and Refrigeration* found that "the average cost of a service call was $3.00 and that, at the best, [the refrigerators] would average at least four a year, some installations requiring several times that number of service calls."[13] The retailers for the largest refrigerator manufacturing firms even had to do expensive preservicing before installing refrigerators.[14] No wonder so many early household refrigerator firms failed.

The need for all these service calls severely damaged the reputation of these machines. The journal *Electrical Merchandising* reported in 1926: "Refrigerator manufacturers in a great many cases have come in for very pointed criticism because units have been shipped out of the factory entirely lacking in final inspection and showing unmistakable evidence of hurried assembly."[15] Ample evidence for this assessment exists in the appendices to Stevenson's memo. General Electric bought, tested, observed, or at least collected reports on all the household refrigerators made by its potential competitors before it entered the market itself. One machine did "not have sufficient area for heat transfer to maintain proper temperatures in the food compartment in a 90 degree room." A fungus that tended to build up on the sulfur dioxide over time slowed the early Kelvinators, since it hampered the distribution of the refrigerant. When the Western Electric Company tested another firm's refrigerator, the sulfur dioxide leaked, and it took two men three days to recharge the machine.[16] As late as 1925, a writer for *Scientific American* noted that while electric refrigerators had been advertised for a number of years, "yet you do not find them in universal use nor are they even commonplace. Although the initial cost of these home units is no higher than that of a flivver, and the upkeep less, they have not yet reached the stage in their evolution where operation is near enough to perfection to bring the wide market which their many good points warrant were they as dependable as, for example, the electric light."[17]

Besides the inconvenience, the need for having a refrigerator repairmen appear (or reappear) at the door created other difficulties. A malfunctioning refrigerator could easily spoil the food it contained. The same thing could happen when ice deliverymen skipped a day, but housewives could always call for more ice the next day. If a refrigerator needed a part not found locally, fixing that appliance could take substantially longer and the household was without any means of cold storage.

As a result of the poor reputation of refrigerators in general, the earliest Frigidaires had a five-year guarantee. If a repairman could not repair a customer's refrigerator, the local office shipped it back to Detroit, at a considerable cost, for service.[18] This explains why the company had every incentive to fix its models in the field, but doing so had its drawbacks. One serviceman, R. O. Ashton, visited the wife of a General Motors official in suburban New Jersey to fix a sulfur dioxide leak. Because she had a bridge game that afternoon, she was particularly concerned that none of the gas leak into other parts of the house. After sealing the doors and opening the kitchen window, Ashton began to fix the problem, but the volume of gas leaking from the refrigerator quickly overwhelmed him. He tripped on the cellar door during his escape, startling a gardener, who disturbed the entire neighborhood with his shouts for help. The bridge party broke up early. "By such adventures," concludes the internal history of the firm in which this story appears, "Frigidaire servicemen acquired strange reputations."[19] Despite such problems, many users remained loyal to the company, undoubtedly because of the incredible convenience that household refrigeration provided when this technology worked.[20] Electrical utilities played upon this sentiment in order to sell refrigerators, which in turn drove demand for their regular product.

The Role of Central Stations

Although he was fully aware of the problems associated with household refrigerators during the early 1920s, Alexander Stevenson recommended that General Electric enter this industry because of the effect that these appliances had upon GE's electric power generating business. "The General Electric Company gains by anything which benefits the electrical industry," Stevenson wrote. "There is no doubt but [that] the electrical household refrigerator is the best household appliance from the standpoint of revenue to the central stations."[21] Since refrigerators always remain on, they use much more electric power than any other appliance. According to a 1924 survey, refrigerators offered central power stations over four times as much revenue per kilowatt hour as any other appliance then on the market.[22] A 1926 study suggested that simply installing an electrical refrigerator doubled the revenue that an electric company received from a home.[23]

Despite such potential benefits, electrical utilities took only tentative steps to enter the household refrigeration business during the early 1920s. A 1924 pamphlet from New York Edison merely lists the refrigerator as one of many appliances available for "electrical housekeeping" so that housewives can have more leisure hours. It gets the same amount of space in the pamphlet as an electric heating pad.[24] This reluctance might have derived from the cost of developing an infrastructure that could support the kinds of household refrigerators then on the market. As late as 1925, the motors in most refrigerators could overload the electric circuits installed in houses at that time. The resulting sudden drop in voltage was a source of annoyance to many homeowners.[25] That alone might explain why only 1 percent of wired homes had electric refrigerators by that year.[26] In 1924, the Electric Domestic Refrigeration Committee of the National Electric Light Association (NELA) noted that "troubles of various kinds have resulted in this apparatus [the refrigerator] proving highly unsatisfactory and concerns handling it being placed in embarrassing positions and suffering heavy losses financially."[27]

Nevertheless, the potential profits to be gained by serving customers with such power-hungry appliances were so great that the utilities changed their minds even before all the glitches in household refrigerators were worked out.[28] By 1926, NELA's Electric Domestic Refrigeration Committee was recommending that public utilities "should be enthusiastic to go ahead and aggressively push the sale of the domestic electric refrigerating machine."[29] In 1931, the NELA began a massive advertising campaign coordinated with seventy-five manufacturers designed to sell one million units in a single year.[30] Their member utilities even offered easy payment plans for potential buyers before the refrigeration makers did.[31] Because central power stations did not depend on the sale of appliances as the source of their profits, they sold refrigerators at cost or even a loss in order to gain the inevitable electrical consumption generated over the life of the appliance.[32]

Marketing by central stations also played a huge role in cultivating new demand for household refrigerators. In the early 1920s, electric refrigerator manufacturers depended upon door-to-door salesmen (like the ones who gave the speech detailed at the beginning of this chapter) to sell their appliances and used little national advertising.[33] Electric utilities, on the other hand, sold refrigerators from impromptu appliance stores that they serviced themselves. They already had experience organizing demonstrations of other electric home appliances and used the same aggressive, large-scale marketing methods that they did when selling other appliances.[34] A 1926 study found that in 154 localities electric utilities sold 31.9 percent of all electric refrigerators. Their endorsement of this technology not only did much to establish customers' trust in these machines; it actually increased sales at other dealers.[35] In this manner, the utilities served as a transitional distributor for manufacturing companies to sell their product until their own distribution networks were established.

Despite the obvious need for electric household refrigerators, the earliest models did not sell well. Advertising for such refrigerators appeared in major magazines as early as 1916, but only 10,000 units of all makes and models had

Table 7.1. Total household refrigerator sales nationwide

Year	Units sold
1910	100
1912	175
1914	600
1916	1,000
1918	1,400
1920	4,000
1922	10,000
1924	17,000
1925	75,000
1926	248,000
1928	468,000

Source: Anne Francis Miksinski, "The Quiet Revolution: The History and Effects of Domestic Refrigeration in America; From Ice to Mechanical Refrigerants," M.A. thesis, George Washington University, 1970, 54.

been sold by 1920. Early manufacturers did not help their own sales by writing operating instructions for these earliest mechanical refrigerators that suggested the many ways in which these appliances still resembled iceboxes. Copeland, an early refrigerator manufacturer, gave its users advice on the best way to arrange foods inside of the appliance: "the bottom of the refrigerator is coldest," and "each higher shelf is a few degrees warmer."[36] With little incentive for consumers to change, sales grew slowly through 1925, until most of the mechanical difficulties associated with this technology disappeared and ease of use significantly improved.[37] By that year, not only had the sales network improved, but the best household refrigerators were quiet, fully automatic, and within the price range of average consumers.[38] Table 7.1 illustrates the cumulative effect on sales of all these improvements in technology and distribution. Higher sales from 1925 on reflect the newfound ability of refrigerator manufacturers to produce their product on a mass scale. However, only a tiny number of firms ever got big enough to make that possible.

The Structure of the Household Refrigerator Industry

The huge potential sales for household refrigerators attracted an enormous number of firms into this industry during the 1910s and 1920s. Many of these companies made fewer than twenty-five units each year.[39] Most of them did not survive. Some of the early refrigerator companies did not even produce a functioning product. They simply depended upon exploiting a gullible public that did not understand how refrigeration worked.[40] In 1932, the journal *Electric Refrigeration News* published a list of extinct household refrigeration makers. By that time, 111 firms had quit the business, 41 could not be located by mail, and 24 had been taken over by other manufacturers.[41] This shows that tremendous barriers to entry existed for the electric household refrigerator market. The firms that survived made use of economies of scale, mass production, and large advertising budgets to turn this appliance from something that only rich people could afford to a common feature of the American kitchen. Even then, it was not until World War II that most Americans owned a refrigerator. Simply surviving this industry's growing pains required a lot of capital. By 1937, the four largest household refrigerator companies controlled 70 percent of the business in the industry.[42]

The power of mass production eventually forced the price of household refrigerators significantly lower. The first machines cost a lot, but the average cost of a household refrigerator dropped far and fast. In 1920, the average cost was $600; by 1930, it was $275; and by 1940, after twenty years of steady mechanical improvements, the average cost of a household refrigerator in America dropped down to $152. (The average cost of operating a refrigerator decreased too.)[43] The rise of a few large household refrigerator companies with the resources for mass production made this drop possible. By significantly improving the quality of household refrigerators and by making them cheap enough so that most Americans could purchase a unit, these firms created an important turning point in the history of the refrigeration industry. The success of these large refrigerator manufacturers limited the growth of smaller competitors that depended upon different technologies than the ones we all know today. The Servel Corporation serves as a good case study of these alternative technologies.

Servel began marketing the first and only gas-powered ammonia absorption home refrigerator in 1926, the same year that the market for home refrigerators began to take off. (Ironically for a company that came to specialize in gas-powered refrigeration, its name stood for "Serve Electrically.") Engineers who reviewed the design of the machine found it "ingenious" and "clever." Customers liked that it was virtually silent and that it had few moving parts, which meant fewer breakdowns. Although the firm initially also made electric refrigerators, it stopped making electric units altogether after a few years and made only gas-powered ammonia absorption machines. As a result, unlike many other refrigerator makers, including more than a few others that did not use electricity as a power source, Servel actually survived the early history of this industry. Compared with its smaller competitors, Servel flourished. Compared with its giant electric refrigerator counterparts, Servel was a small fish in a big pond. The company never had as much capital as competitors who relied on compression for their refrigerators. Servel continued to build refrigerators until 1956, but it never obtained more than an 8 to 10 percent share of the home refrigerator market.[44] Servel's story suggests one reason why compression machines eventually beat out absorption machines in the American home refrigeration: the resources of the refrigerating manufacturers who produced compression machines.

Despite the dominance of compression refrigerator manufacturers during the late 1920s, it still seemed possible to break into the household refrigeration industry at that time, and a few individual mechanics tried to start their own companies. A very small number of firms with large backers carved out niches during the 1930s (an example was Sears Roebuck's Coldspot, first introduced in 1928), but most startups failed because the mass producers undercut them on price.[45] To make matters worse, these powerful, aggressive companies also spent a fortune on advertising to establish their brands. The industry's total advertising expenditures grew from only $45,000 in 1923 to $20 million in 1931.[46] It took plenty of capital, excellent engineering talent, and a good sales force to survive the early days of this industry.

The Guardian Refrigerator Company of Detroit survived in the sense that it was absorbed by a larger concern. Alfred Mellowes and Reuben Bechtold founded Guardian in 1916. Their refrigerators came in a sturdy oak cabinet, cost $700 retail, and offered only nine cubic feet of food storage space. The company produced fewer than forty such appliances and had lost $40,000 when General Motors president William Durant bought the bankrupt firm in 1918 and renamed it Frigidaire. Durant's insistence on applying the mass production techniques of the automobile industry to refrigerators explains the company's turnaround. The troubles with early models hurt sales, but engineering improvements and aggressive marketing quickly turned Frigidaires into a popular product. In 1921, Durant moved the company to Dayton, Ohio, where it became part of the Delco-Light Company division of General Motors. By 1923, the profits from Frigidaire covered the entire dividend of GM's preferred stock.[47] Frigidaire produced its millionth refrigerator in 1929 and had more than doubled that number three years later, having produced 2.25 million refrigerators by 1932.[48]

Like Frigidaire, Kelvinator had its origins in the auto industry. Edmund Copeland, formerly of General Motors, and a broker named Arnold Goss began experimenting with marketable household refrigerator designs in 1914. They tested their first compression model in 1917 and founded Kelvinator the next year.[49] They named the firm after Lord Kelvin, the Scottish physicist who developed the basis for absolute zero and whose name is given to the scale for absolute temperature.[50] The first Kelvinators resembled the Isko. Customers bought an icebox for the kitchen. Kelvinator installed its condensing unit in the basement, and connected the two parts with a series of pipes that went through a hole in the floor. Early models did not control temperature well, and the sulfur dioxide refrigerant often leaked, as so many other early household refrigerants did.[51] Kelvinator did not combine the machinery and the refrigerator into one unit until 1925.[52] That year, the company sold 75,000 units of its new combined models; and sales skyrocketed after that, to 850,000 in 1930 and 2.5 million by 1937.[53]

General Electric took a more measured approach to electric household refrigeration than its competitors from the auto industry. Its first attempt to enter the market came by acquiring the American patent rights to the Audiffren, a French refrigerator that used sulfur dioxide as the refrigerant, in 1911. Because of the machine's high price, GE never sold more than a few hundred of them in any given year.[54] The company began its own research on electric household refrigerators in 1917, but did not seriously consider mass-producing a refrigerator of its own until the issuance of Stevenson's memo in 1923. The company used the Audiffren as the basis to develop something better. In 1926, it sold as many as 2,000 of its OC-2 refrigerating units through electric utilities. These units, like the earlier Isko and Kelvinator models, fit inside an icebox. The OC-2 units were placed inside GE's own icebox, which the company contracted from the Seeger Refrigerator Company of St. Paul. (GE would not fit its machines into other iceboxes.) Like the Isko and Kelvinator before it, the OC-2 also had service problems. Because of the varying quality of available iceboxes, GE really wanted to produce its own stand-alone refrigerator.[55]

By waiting until 1927 to release its first modern household refrigerator, General Electric learned many lessons from the unsuccessful attempts of smaller firms to develop a similar machine. "We did not . . . go into production until the research had been *finished*," the company explained in an ad to get distribu-

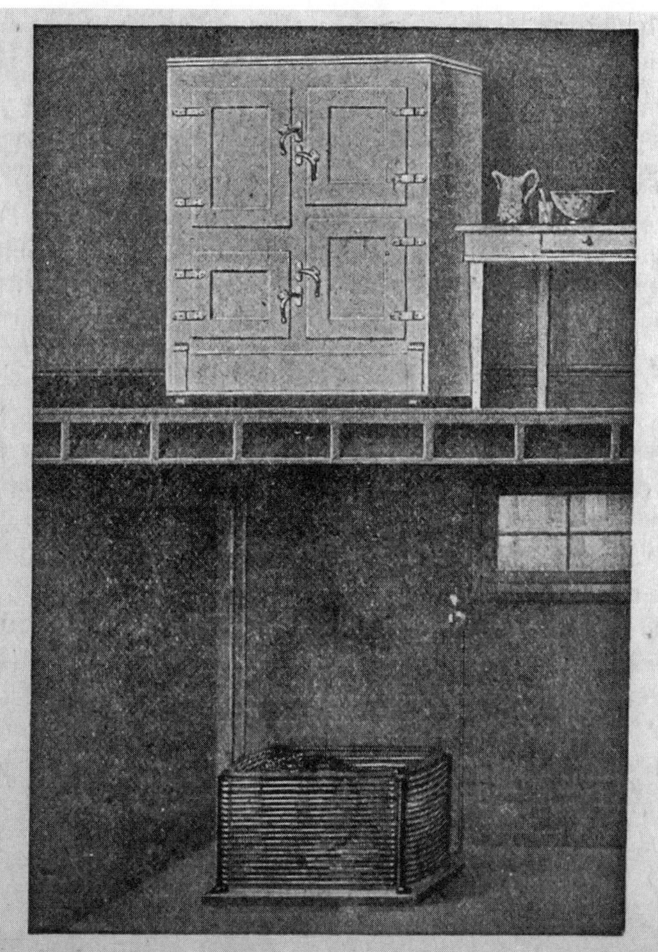

Illustration showing domestic refrigerating plant installed in existing ice box. It is entirely automatic in operation. Made by the Kelvinator Corp., Detroit, Mich.

The early Kelvinator refrigerator required owners to cut a hole in their kitchen floor so that the condensing machinery could connect to the appliance by means of a belt. *Electrical Record*, Oct. 1919, 207. Courtesy of the Hagley Museum and Library.

tors to sell the new product. "To have done so [otherwise] would have entailed great risk. A great risk for ourselves, our dealers, and our customers."[56] Its sales literature explained the objective of its research even more clearly. GE had created "an electric refrigerator so efficient, so simplified, that it could literally be installed and forgotten . . . a refrigerator so simple that all you need to do is to plug it into the nearest electric outlet and it never even needs oiling." The intended contrast with earlier household refrigerators or iceboxes could not have been greater. "There isn't a single belt, fan or drain pipe—nothing below the cabinet—nothing in the basement." GE's new refrigerator came in eight different models, ranging from two and a half to eighteen cubic feet, in order to fit in any size kitchen.[57]

GE called this new refrigerator the "Monitor Top" because its top reminded one company wag of the turret of the Civil War battleship *Monitor*. The sealing on the compressor had to be airtight so that the noxious sulfur dioxide refrigerant could not escape.[58] This made the machine run significantly more quietly and decreased the probability of oil leaks.[59] "There is not a single exposed moving part," the company bragged. "All the mechanism is hermetically sealed in a steel casing on top of the cabinet. There are no troublesome belts, fans, stuffing boxes or drain pipes."[60] These advanced features and the model's low price made it a stunning success; by 1929, GE had sold 50,000 of these refrigerators and had sold a million by 1931. Improved versions of the Monitor Top continued to be sold for the next ten years; they are prized by antiques collectors, as many still work today.[61]

The economies of scale that made mass production of refrigerators like the Monitor Top profitable depended upon a mass market. Without low prices, that mass market never would have existed. Refrigerators were still a comparatively expensive appliance at the end of the 1920s. Keeping the prices of mass-produced units low compared with other refrigerators made it possible for middle-class and even some lower-class customers to afford them. Had these refrigerators remained unreliable, lower prices alone would not have kept driving sales during the Depression decade of the 1930s, but refrigerators practically sold themselves during that difficult economic time. The continued appeal of household refrigerators explains why refrigerator manufacturers and individual retailers undertook huge advertising campaigns after 1925, spending on average between 5 and 10 percent of their total sales.[62] Much of that advertising centered on the inadequacy of the refrigeration technology that came before electric refrigerators, namely, the ice industry.

Getting Rid of the Iceman

In the 1932 Mack Sennett comedy *The Dentist*, W. C. Fields plays a dentist who gets upset when he finds out that his daughter intends to marry an iceman. "He goes to college. He's a Cornell man," the daughter argues, but Fields's character will have none of that. When she informs him that her fiancé will have to return to their house soon with another fifty-pound block for the icebox, he replies: "Keep that iceman out of here. I'm going to order a Frigidaire." After his daughter elopes with the iceman at the end of the film, she asks him, "Father, you're not really going to buy a Frigidaire, are you?" He responds, "Fifty pounds and make it snappy," to connote his approval. Ten years earlier, consumers offered much greater resistance to the idea of getting rid of the iceman. Refrigerator manufacturers had to show people why they should remove an icebox that still worked and buy a much more expensive product. To overcome buyer resistance, the mechanical refrigeration industry used its advertising to portray the disadvantages of its predecessor technology in very unflattering terms.

The arguments used by the mechanical refrigeration industry against the ice industry strongly echoed those in favor of mechanical ice over natural ice, and even those touting the advantages of expensive iceboxes over cheaper ones. Servel, in a 1925 ad for an electric refrigerator that predated its gas unit, proclaimed: "Ice—the best food keeping method of past years—adds to the duties of the home-maker's busy day. But a Servel electric refrigerator is as labor-saving as an electric iron, washing machine or telephone." The electric refrigerator, the ad said, is "the gift that strikes off the shackles of ice."[63] Other firms took direct shots of the ice industry's most public face: "The ice man," the first issue of *Frigidaire* magazine said, "may be a good and well-meaning fellow, but there are many reasons he has got to go. Here are three: He does not keep clean-nor-quiet—NOR ON TIME."[64]

That slogan serves as a good guide to most of the common critiques of the ice industry in its later years. For example, the argument about the iceman being unclean echoed earlier ads by the makers of expensive iceboxes that icemen could service without entering the house. A long, but well-written piece of such prose sent by a Kelvinator sales representative to a hotelier in 1921 makes this argument well:

> Did you ever follow a block of ice from its source? It is dragged across the floor of the ice house and dumped into a truck. Here exposed to the air, dust, flies and germs, it is hauled all over town. By its very nature it is the perfect germ catcher. A sheet of fly paper couldn't be better.
>
> The driver adjusts his horses' harness, or, examines their feet, or he may tinker with his engine. Then after a chew or a smoke he arrives at your home. Bang! goes

Some expensive iceboxes like the McCray Patent Refrigerator had doors outside the house so that the iceman would not track dirt into customers' houses. This feature also allowed customers to get their ice replenished even if they were not at home. JWT Domestic Advertising Collection, Box 1, Ad Portfolio—Database #J0015, Emergence of Advertising Project, John W. Hartman Center for Sales, Advertising & Marketing History. Courtesy of Duke University David M. Rubenstein Rare Book & Manuscript Library, http://library.duke.edu/digitalcollections/eaa.

> the ice on the curb while he gets a fresh grip. Lugging ice is a heavy job. The hotter the day the harder the work and the more the driver perspires.
>
> Now watch him get the ice onto his shoulder, against his shirt or an old burlap sack. Here he comes. If you're not spry—he is obliging and will move your food for you—and in goes the ice. Pleasant, isn't it?
>
> As to whether that ice is clean, examine your drain pipe. You've found your answer.[65]

The shorter the cold chain, the more you could have faith in the product's cleanliness, and no shorter cold chain existed than making ice in your own kitchen. This argument fit into a general campaign to suggest that living without a household refrigerator could be dangerous. "Forty-five percent of all deaths are caused by diseases of the stomach and intestines," Kelvinator claimed in 1924, citing the *Journal of Medical Sciences* as its source.[66] "The importance of correct food preservation as provided by Frigidaire cannot be over-emphasized," that company asserted in a 1927 sales pamphlet. "Too many illnesses have been traced to improperly kept food to side-step this vital point."[67] These kinds of arguments served double duty, reminding potential buyers of the inadequacy of ice and the benefits of reliable mechanical refrigeration at the same time.

The second complaint, about the iceman not being quiet, referred to the distraction of having someone knock upon your door and then carry a large block of ice through your living room on the way to the kitchen. It is the same problem that resulted in outside doors on expensive refrigerators earlier in the century. Electric refrigeration companies began to claim that household refrigerators ran quietly as early as Isko, around 1918.[68] With the main source of noise located in the basement, this argument made sense. Other kinds of household refrigeration units still had a serious noise problem well into the 1920s. Writing about a Frigidaire in 1922, a reporter for the *New York Tribune* noted that "the hum of the motor and purr of gears, while not at all unpleasant, may be considerably muffled by seeing to it that the part of the floor on which the cabinet is to set is perfectly solid."[69] General Electric solved the noise problem for all time by immersing the moving parts of the compressor of its Monitor Top in a permanent bath of oil.[70]

Frigidaire's third critique—that the iceman was always late—invoked the problem of finding good delivery as well as the common complaint that late delivery led to spoiled food. Another Frigidaire booklet repeated this criticism: "Waiting for ice . . . is a further annoyance that is often caused by the fact that some member of the household wants to leave, but cannot because the ice man has not come." In contrast, "no one has to stay home when you own a Frigidaire. While you are away, Frigidaire works automatically and makes ice which is always ready for use. It is very convenient to be able to leave at the week-end or for a vacation, with no worry about food spoiling, ice delivery, or other matters pertaining to the refrigerator."[71] The ice industry could never hope to match electric appliances in the convenience department.

Despite the initial costs of refrigerators, refrigerator manufacturers argued that paying the iceman still cost more because iceboxes required customers to buy ice forever. Kelvinator "saves ice bills," that company argued. "You pay the ice man as long as you live. Buy a KELVINATOR and you finish paying."[72] For a public that was used to continually shifting ice prices, the idea that their household finances could become more stable after purchasing a household refrigerator must have been very appealing. Consumers also valued the convenience that refrigerators provided, particularly during the Bull Market years of the 1920s.

In an early recipe book entitled *For the Hostess*, Kelvinator summed up the benefits of its product this way: "The housewife sees her labors lightened, sees more hours of leisure, and with it all, extraordinary economies."[73] For example, refrigerators decreased the frequency with which homemakers had to go shopping. "Once I had to purchase food nearly every day," explained a testimonial in a 1926 Servel pamphlet. "Now I rarely go to the market oftener than twice a week; and because I buy in larger quantities I purchase more cheaply."[74] Household refrigerators also allowed homemakers to prepare pieces of large meals in advance and take them out when company came. If you had advance warning of company's arrival (as at Thanksgiving), you could prepare all the exotic chilled dishes you wanted to serve a day ahead of time.

To get rid of the iceman once and for all, you had to be able to make ice cubes in your new electric refrigerator. While technically, freezers did not exist in early models of household refrigerators, the area next to the compressor was cold enough to freeze water. The first ice cube tray came as part of the DOMELRE, and other manufacturers subsequently copied this innovation.[75] The ability to make ice cubes in your own kitchen became an incredibly effective selling point for electric household refrigerators during the 1920s. What seems so mundane now came up frequently in the marketing literature, described in language that seems bizarre in retrospect. "Ice cubes frozen in the new Kelvinator are instantly available," explained a pamphlet devoted entirely to this subject. "You merely bend the rubber tray. The cubes slip out twinkling. Imagine the joy of filling a tray at your own kitchen sink, placing it in the Kelvinator, and in just a few hours—convenient ice cubes for all purposes!"[76] Recipe books issued by refrigerator companies recommended freezing maraschino cherries or sprigs of mint into your ice cubes in order to make them extra special.

The natural ice industry did very little during the 1920s to counter this onslaught of negative publicity against it and positive publicity for its competitor. The industry's trade association, the National Association of Ice Industries (NAII), believed that the low cost of ice as compared with the cost of electric refrigeration would persist indefinitely.[77] It also downplayed the significance of the competition because of the technical flaws then present in household refrigerators. At a convention reported on in *Ice and Refrigeration* in 1924, one executive stated, "From the best information we have been able to secure, we do not believe that the household refrigeration machine can compete with ice on a basis of relative economy." He also told his audience, "You know the American people are prone to take up a fad and run away with it."[78] In November 1926, a New York banker told the Eastern Ice Association, "There is a very definite feeling in the last 60-day period that the mechanical refrigerator is not making the headway that was prophesied for it early this year."[79]

Some individual ice companies did conduct aggressive anti-refrigerator advertising campaigns. "*You* were never supposed to see this," explained an ad for an Ohio ice company. "We publish it to settle, once and for all, the question of the cost of electric refrigeration." The ad featured a statement from an electric power representative noting that one electric refrigerator earned the same revenue as fifty electric fans, twenty-nine vacuum cleaners, or eleven electric irons, a sentiment that utility industry representatives often expressed publicly.[80] Another ice company ad read: "No improvement has been made upon the convenience of ice. [It] entails no costly machinery, no charge for electric current, no repair bills, no noise. There are no fuses to blow nor belts to break—nothing to worry about when you are out of the city during the week-end."[81] While ice still had some advantages over electric refrigeration in 1920, they had all disappeared by the end of the decade because of the dramatic improvements in the new appliances.

Ironically, all the advertising for refrigeration of any kind actually increased the sale of ice during the early 1920s. People unable to afford the convenience of electric refrigeration bought more ice, since it could still keep their food fresh and healthful.[82] Because of this publicity, the total number of ice plants in the United States increased from 5,117 with an output of 28.2 million tons of ice per year in 1920, to 7,338 with an output of 43.5 million tons in 1928.[83] Since both sides of this struggle found economic success, industry representatives made repeated calls for a ceasefire in the war of rhetoric between these two kinds of cold providers so that both parties could reap the rewards. While that might have stemmed the tide briefly, a campaign of education and improved marketing could only take the ice industry so far. As electric household refrigerators became more common, the public gradually came to refer to both iceboxes and electrical refrigerators with the single term "refrigerator." While this may not seem significant in retrospect, it nonetheless demonstrates that a considerable amount of the advertising touting the unique qualities of ice refrigeration did not penetrate the American consciousness.[84] People used the word for both appliances because they cared more about the cold than how that cold got generated.

Unlike the competition between gas absorption and electric compression home refrigerators, the competition between ice and electric refrigeration was on decidedly unequal grounds, since ice offered significantly inferior refrigeration to the electric household unit. If it had not been for the difficulties early in the development of mechanical refrigeration and the inability of some Americans to afford that alternative, the ice industry would have disappeared long before the Great Depression. But rather than addressing its long-term problems, the ice industry liked to wax eloquent about the great untapped market for new ice customers. A 1926 conference paper summarized in *Ice and Refrigeration* reported that just 54 percent of American families used any kind of refrigeration, and only 15 percent used refrigeration of any kind during the winter.[85] Yet three years earlier, a much more realistic ice industry executive admitted, "There is no doubt about the fact that a great many women have been dissatisfied with the refrigeration they get from the ice they use."[86] Once electric household refrigerators improved and their cost dropped, the notion that ice could thrive, let alone survive in the future was absurd since it was so ineffective by comparison. Cost alone could not keep iceboxes a fixture in the American home.

Refrigerants for the Household

On 3 June 1929, gas fumes leaking from their household refrigerator overcame the entire Irving Markowski family of Chicago, Illinois. Mr. and Mrs. Markowski and three of their children survived, but the Markowskis' three other small children died. Later that same month, Mrs. Violet Clarke of Chicago died as result of a leak from her household refrigerator. Her husband and mother survived but were hospitalized in critical condition. As a result of these tragedies, Chicago's health commissioner investigated other similar incidents and found that fifteen people had been made ill by leaking refrigerators and seven had died in just the previous few months. Dr. Morris Fishbein, the editor of the *Journal of the American Medical Association*, headed a coroner's jury that investigated Mrs. Clarke's death. That group concluded that "although hundreds of thousands of refrigerators have been installed all over the country and they operate with apparent slight mortality . . . the public should be properly protected. . . . Both public officials and heads in the refrigerating industry must take steps to prevent further sickness and death from these causes."[87]

All of these deaths had been from suffocation caused by methyl chloride, the refrigerant used in each of these leaking refrigerators. While these tragedies could have led to devastating ads attacking the safety of methyl chloride, refrigerator manufacturers had an informal agreement not to attack each other on the basis of the potentially fatal properties of their refrigerants. The newspapers, however, showed no such restraint.[88] "Methyl chloride is insidious," exclaimed one Chicago paper during the wave of deaths. "The victims neither smell nor perceive the deadly gas until they become physically inert and mentally fogged."[89] Victims of methyl chloride poisoning feel extremely hot, so in this wave of tragedies safety officials discovered the deceased either nearly or totally naked.[90] One refrigerator maker claimed that "faulty installation" caused all such poisoning deaths, but these excuses proved unconvincing to a public already predisposed to think of mechanical refrigerating equipment as dangerous.[91]

The refrigeration industry in America and elsewhere had a long history dealing with refrigerant leaks, going back before the Columbian Exposition disaster in 1892. Ammonia often leaked from large-scale refrigeration plants (although not all ammonia leaks led to explosions). A story about one such leak in Germany convinced Szilard and Einstein to experiment with their electromagnetically powered model.[92] Thanks to years of faulty media coverage of explosions (many of them incorrectly attributed to ammonia), refrigerator manufacturers knew that ammonia would not gain acceptance as a refrigerant inside the American home. Although it continued as the refrigerant of choice for industry (and remains so today), many new (or at least rarely used) refrigerants were chosen for early household refrigerator models.

Compared with the millions of household refrigerators sold in the United States during the late 1920s and early 1930s, the number of deaths caused by refrigerant leaks were actually minimal. But because the use of household refrigerators marked the first time that refrigeration equipment appeared inside the home, safety became a prime concern for manufacturers.[93] Most of the early refrigerators used sulfur dioxide as the refrigerant because it was the safest, most practical option available.[94] Even though the gas is poisonous, its noxious rotten-egg smell meant that most refrigerator owners could evacuate their homes in the event of a leak before getting poisoned. Unfortunately, if water gets into a refrigerator's closed compression system, it bonds with the sulfur dioxide to form sulfuric acid, which will eat through whatever part of the machinery it touches.[95] Other substances used as refrigerants included ethyl chloride, butane, propane, ethane, and ether. Some inventors created household refrigerators based on the absorption system used in the larger multi-ton units of the previous century, in which aqua ammonia was the refrigerant. One of these, the Icy-Ball, first appeared on the market in 1927. The diagram that appears in Althouse and

Turnquist's *Modern Electric and Gas Refrigeration* makes it look like a Japanese lantern hooked up to a kerosene stove, which was, in fact, the heat source used to power its refrigeration cycle.[96] This model required a lot of tending: its cycle lasted only twenty-four to thirty-six hours, at which point the apparatus had to be removed and turned upside-down in order for the refrigerant to return to its original position.[97]

The existence of such a contraption on the market demonstrates one thing clearly: the inability of manufacturers to settle on a single refrigerant for household models left an opening for a refrigerant like methyl chloride, which despite its safety problems actually had many advantages. It had been manufactured in Europe since 1875, and starting in 1920, the chemical producer Roessler and Hasslacher began to manufacture and market methyl chloride as a "safer" alternative to ammonia for home refrigerators. While this substance is flammable and can dissolve rubber joints, it still makes an efficient refrigerant, almost as efficient as ammonia.[98] Methyl chloride became especially popular in apartment houses, where landlords often installed many refrigerators at once. While it can cause death through suffocation, methyl chloride is actually seventy times less toxic than sulfur dioxide.[99] Nevertheless, the poor reputation that this refrigerant acquired after the poisonings in Chicago encouraged forward-looking manufacturers to look for a new refrigerant which could put all safety concerns about household models to rest.

Willis Carrier and his associates started experiments to find a new refrigerant for air-conditioners in 1918 because of the safety hazards associated with ammonia and sulfur dioxide.[100] Larger manufacturers of small refrigerating units also began to look for a safer refrigerant shortly thereafter. Du Pont bought Roessler and Hasslacher in 1930 despite the dangers of the refrigerants that firm produced because it wanted another one of its products. In 1928, General Motors commissioned Thomas Midgely, the inventor of leaded gasoline, to develop a new refrigerant for household units at the GM Research Laboratories in Dayton. Midgely pursued the development of chlorine- and fluorine-substituted hydrocarbons (chlorofluorocarbons, or CFCs) because they were nontoxic, nonflammable, and stable. None of these compounds had ever existed before in nature.[101]

Frigidaire took out a patent on CFCs in 1930. That same year, GM (Frigidaire's parent company) created a joint venture with Du Pont, the Kinetic Chemicals Company, to manufacture and market CFCs, and to further develop them for a variety of uses.[102] The Patent Office approved the patent in 1931, at which time the companies introduced these new refrigerants to the public. The companies named all the chemicals in the new line Freon. The most popular was Freon-12.[103] Initially, Freon-12 cost $50 per pound to produce. Sulfur dioxide cost only six cents per pound at this time. In order to promote the acceptance of Freon, Du Pont nonetheless sold it at a dollar per pound despite the potential to cannibalize the market for other Du Pont refrigerants.[104] By the end of the 1930s this proved a wise business decision, as the Freons made the company much more money.[105] By the 1970s, Freon had supplanted all the old refrigerants except for ammonia, which remained the refrigerant of choice for industrial installations.[106] Safety concerns explain why Freon became so popular so quickly. A safe, reliable refrigerant for home installations meant that the cold chain, from production all the way through home refrigeration, had almost come of age. The only remaining challenge was a reliable mechanically refrigerated railway car.

With electric household refrigeration now reliable, compact, precise, and affordable, this technology did not improve in future years so much as expand. Some technological changes occurred along the cold chain after 1930, but they did not change the basic fact that people had conquered the cold for their own ends. As refrigerators got bigger and freezers became common, the achievement of the cold chain itself was increasingly forgotten in the United States; and defying geography and season became affordable and commonplace. In recent decades, cold chains have propagated and lengthened, moving more goods and including more countries. Like ice in the nineteenth century, modern refrigeration serves a dual function as both industrial infrastructure and a common part of everyday life. As Americans have become accustomed to cheap and convenient food, refrigeration and freezing have become increasingly important to even the poorest Americans.

CHAPTER 8

The Completion of the Modern Cold Chain

"When a housewife returns from the supermarket and whisks things into her refrigerator and closes the door," wrote Kathleen Ann Smallzried in 1956, "she has closed the door on the springhouse, the milk and butter pantry, the root cellar, the cheese room, the smokehouse, the covered well. At the same time she has turned her back on the preserving kettle, the pickling crock, the pudding bag, the vinegar barrel. As for the icehouse [and] the ice wagon, she has put them behind her too. Ice does not make her storage box cold. Instead, the box makes ice for all her needs."[1] The advent of the global cold chain coincided with post–World War II prosperity in the United States. Now that American families had a place to store the bounty that modern agriculture could produce, nothing could stop them from spending their newfound wealth on perishable food imported from all over the world. Americans also spent a portion of their new wealth on bigger refrigerators. As their refrigerators grew larger, Americans purchased newly available fresh and packaged perishable foods in just about any quantities they wanted, since they now had more refrigerator capacity than ever before. During the nineteenth century, America's embrace of refrigeration came about as a by-product of its natural abundance. Now the country's growing dependence upon refrigeration served as a sign of its extraordinary wealth.

During the Depression, many people could not afford household refrigerators and had to keep buying ice. Therefore, the iceman survived well into the post–World War II period. After the war, the ice industry still thrived in the cities where it had originated a century before. "No matter where you live in New York," wrote E. B. White in 1949, "you will find within a block or two . . . an ice-coal-and-wood cellar (where you write your order on a pad outside as you walk by)."[2] Since 30 percent of American households still had iceboxes in 1947 (sometimes in conjunction with an electric refrigerator), the Vivian Manufacturing Company of St. Louis still supplied ice companies with all the supplies they needed—including personalized ice picks stamped "Those Who Really Know Prefer Ice" to hand out to customers.[3] Some people still did prefer ice, but not enough to keep the industry going for very long. The ice industry in its original form probably went extinct sometime during the mid-1950s.

The last days of the ice industry looked nothing like its heyday. Despite its survival, the rise of household refrigerators had a brutal effect upon ice profits. As the quality of electric home refrigerators improved and their price dropped, the total value of manufactured ice in the United States dropped from $211 million in 1929 to $128 million in 1935.[4] In 1930, total electric household refrigerator sales exceeded those of iceboxes for the first time.[5] Most American icebox manufacturers went out of business during the 1930s. By 1935, there were 1.7 million electric household refrigerators and only 350,000 remaining iceboxes.[6] As late as 1940, the great bulk of refrigerator sales still went to people who had never owned a mechanical refrigerator before.[7] This suggests that the failure of Americans to take to mechanical refrigeration even faster had more to do with circumstances created by the Great Depression than with a lasting attachment to their iceboxes. American consumers understood the limitations of what ice could do for them and bought mechanical refrigerators in large numbers as soon as they could afford them.

Since its beginning, the modern cold chain has offered consumers variety, affordability, and convenience. The introduction of home freezers magnified these advantages, increasing the length of time that food could be preserved. Because of these advantages, consumers in other countries began to develop the same interest in large refrigerators that Americans had had for much longer. That explains why so many new strands of the cold chain have appeared around the globe in recent years. To understand what the future might hold with respect to the use of refrigerators around the world, one must first understand how refrigerators went from common to ubiquitous in the United States. It then becomes possible to explore their movement into other countries.

The Stages of Refrigerator Adoption

Americans adopted electric household refrigerators much faster than a host of other comparable household technologies, including the clothes washer and the color television.[8] It took twenty years for refrigerators to become widespread in the United States. Americans took to few other technologies so quickly. Sales turned upward when the performance of the units improved, especially after 1925. As time passed, less obvious factors played an important role in causing more consumers buy refrigerators. During the period between 1925 and 1945,

Table 8.1. Increase in number of electric household refrigerators, 1925–1941

Year	Est. no. of domestic refrigerators, Jan. 1	% of wired homes with refrigerators
1925	150,000	1.1
1928	755,000	4.3
1929	1,223,000	6.4
1930	1,850,000	9.4
1931	2,610,000	12.8
1932	3,499,000	17.1
1933	4,300,000	21.6
1934	4,900,000	24.6
1935	6,020,000	29.3
1936	7,250,000	34.2
1937	9,000,000	41.1
1938	11,271,000	50.6
1939	12,101,000	51.7
1940	13,701,000	56.0
1941	16,100,000	63.0

Source: Neil H. Borden, *The Economic Effects of Advertising* (Chicago: Richard D. Irwin, 1947), 398.

government played a big part in the growing love affair between Americans and their household refrigerators. Despite low prices, increasing reliability, and heavy advertising, the mass adoption of refrigerators required an important assist from the Roosevelt administration.

Table 8.1 charts the growth of electric household refrigerators between 1925 and 1941 in terms of both the total number of appliances in operation and the percentage of homes wired for electricity. The point at which the majority of American households had electric refrigerators for the first time came at some point during World War II. By the 1950s, just about every household in America had its own refrigerator.

Once the price of electric refrigerators began to decrease, America went through two stages of household refrigerator adoption. The first of these stages came during the 1920s. Refrigerators remained signs of affluence through the end of that decade, even after their reliability increased and prices came down, but they did grow much more common. In Riverside, California, for example, a clear division existed between renters and homeowners. This division illustrates the limits of the refrigerator's penetration. Renters in Riverside, unwilling to make an investment in a kitchen they might soon leave, did not buy refrigerators.[9] Only homeowners did. During the first half of the 1930s, 40 percent of the town's upper class owned a household refrigerator, compared with 20 percent of middle-class households and only 8 percent of working-class households. These numbers mirrored those in other places across the country. In Austin, Texas, only 15 percent of families had electrical refrigerators in 1935, but half of all families with a household income of $3,000 or more had them.[10] This limited adoption rate for refrigerators as late as 1935 came despite a drop of average retail prices from $275 in 1930 to $170 in 1935.[11] This shows that lower price alone could not convince people to buy refrigerators. Considering the state of the economy during the Depression, that makes sense.

The next stage of refrigerator adoption came after 1935, once government loans for appliance purchases became widely available. The National Housing Act of 1934 included provisions for the government to make loans for household modernization, which could include the purchase of appliances. New Dealers and the industry alike promoted electric household refrigerators because they saved money by preventing waste and led to improved sanitation. When these loans were combined with the price decreases in mechanical refrigerators made possible by mass production, mechanical refrigeration became affordable to most Americans despite the Depression.[12] Also, New Deal standards that dictated how consumers could use government loans for appliances encouraged the creation of smaller, nonstandard refrigerators for people who could not have afforded to buy one otherwise.[13]

With loans from the government and an improving economy, the 1935 to 1940 period saw mass adoption of the household refrigerator in Riverside for the first time. While the percentage of upper-class households with refrigerators merely doubled, the percentage of middle-class households with a refrigerator tripled (from 20 to 63 percent), and the percentage of working-class people with refrigerators increased by 285 percent in the second half of the decade over the first. During this second stage, even renters bought home refrigerators because of their value and convenience.[14] These machines worked even better than those built during the late 1920s. In a telling comparison, *House and Garden* remarked in 1935 that "a kitchen equipped today with a refrigerator more than five years old is in a class with a person driving a Model T Ford."[15] Because of the improvements in the technology, many consumers had bought their second refrigerator by the end of the 1930s.

Apart from loans, the New Deal in general also spurred the spread of electric household refrigeration. All forms of government aid better enabled people affected by the Depression to buy new appliances, especially refrigerators.[16] Besides that indirect stimulus, the Roosevelt Administration's programs contributed directly to refrigerator sales. The creation of the Tennessee Valley Authority (TVA) in 1933 made it possible for homes that had once been without electricity to get the electricity they needed to run this most useful of appliances. TVA convinced General Electric to design a cheap, stripped-down refrigerator to sell to TVA's mostly poor rural customers. This government-sponsored experiment in cheap power and cheap refrigerators taught refrigerator makers how to reach the Americans who had not yet taken advantage of what their appliance had to offer.[17] A new influx of refrigerator manufacturers such as Crosley, Norge, and Philco also contributed to refrigerator sales in this period by increasing the supply of models, which put further downward pressure on prices.[18] Similarly, Kelvinator created its first stripped-down household refrigerator in 1940. Other companies followed suit, and the stimulus provided by lower prices and the money that war-related jobs put in people's pockets led to sales of 2.6 million models that year, an all-time high.[19]

Despite limitations on refrigerator manufacturing during the war years (which led to a temporary revival in mechanical ice production), the use of mechanical household refrigeration grew during the conflict because of a shortage of tin plate, which made it harder to buy canned food. By 1944, 85 percent of American homes had refrigerators.[20] This market saturation happened even before the postwar prosperity that ushered in the purchase of so many other types of appliances and demonstrates the popularity of a technology that changed the way its purchasers lived. The electric household refrigerator symbolized modernity. When filled with food, it symbolized abundance. The size of the typical American refrigerator also indicated the extraordinary prosperity of the postwar period.

How the American Refrigerator Got So Big

In his thinly veiled autobiographical masterpiece, *Goodbye Columbus*, Philip Roth writes of a tall, old refrigerator in the Patamkin family basement in northern New Jersey. "This same refrigerator," his narrator observes, "had once stood in the kitchen of an apartment in some four-family house, probably in the same neighborhood where I had lived all my life, first with my parents and then, when the two of them went wheezing off to Arizona, with my aunt and uncle.

After Pearl Harbor, the refrigerator had made the move to Short Hills." Despite its age, it still held a huge bounty of fruit in the basement until the Patamkins brought that food upstairs.[21] The author David Owen tells a similar story of his family's first refrigerator, purchased in 1954, which was moved down to the basement when they bought a new one during the 1960s. There "it remained plugged in for a further twenty-five years—mostly as a warehouse for beverages and leftovers—and where it was soon joined by a stand-alone freezer."[22] The American refrigerator industry wanted Americans to buy big refrigerators, but the Potamkins and the Owens had no need for the biggest refrigerator available as long as they could store their surplus food elsewhere. This presented a problem for the manufacturers: they made their appliances too reliable for their own good.

In 1920, the typical mechanical refrigerator lasted six years. A 1938 model, by contrast, lasted fifteen years.[23] In 1939, General Electric noted that the household refrigerator "has passed beyond the experimental stage. Its growing pains are over. Mechanical refinements which can only come from years and years of actual experience have been incorporated in the G-E mechanism to produce longer mechanical life and lower operating costs."[24] This common trend among all refrigerator manufacturers persisted over time. In 1964, Frigidaire noted in an internal report that "1961 appliances are requiring about nine per cent less service attention than 1960 models. . . . The 1960 model was about 33 per cent better in this respect than 1957, and that 1957 model needed considerably less service than those of ten years ago."[25] By 1988, the typical refrigerator in an American home lasted seventeen years. That figure did not necessarily signify that the old refrigerator had failed, only that the owners chose to buy a new model for whatever reason.[26] The increased use of plastic for parts of the refrigerator helped keep refrigerators running longer. By the 1950s, every new refrigerator had door liners, breaker strips, evaporator doors, ice trays, and control knobs made from some twelve different types of plastic. These components not only lasted longer than the older metal versions of the same thing; they cost less to produce. Such savings helped bring down the price of the entire unit.[27]

As long as old refrigerators worked well enough to keep food fresh, customers had little reason to replace them. This left refrigerator manufacturers emphasizing minor changes in their promotional literature in order to expand sales. "Major" features in the 1937 Servel line, for example, included "push-pull" door latches so that women could open the refrigerator door more easily, hydrators (special trays for vegetable storage), egg containers, and removable

In 1964, Kelvinator promised that growing children with "hollow legs" would "love Foodarama living" to emphasize the huge capacity of its Foodarama refrigerator for middle class families.

split shelves. None of these accoutrements affected the basic underlying technology.[28] "Refrigerators and freezers have changed little in their basic design in decades," the *New York Times* reported in 1992. "Shoppers have grown to expect them to be exceptionally durable and trouble free. Instead manufacturers have competed on the basis of new features like through-the-door ice water and special storage bins."[29] While this kind of sales pitch worked for cars during the 1950s, the refrigerator market could only hope that the cumulative effect of many smaller changes would convince customers to trade in a refrigerator that worked before it ceased to do so.

Refrigerator manufacturers introduced just enough important changes after 1930 that a model from 1970 appeared to differ greatly from a model produced forty years earlier. Frigidaire, for example, introduced the first sealed rotary compressor in 1933, matching color exteriors for its entire appliance line in 1954, modular appliances in 1956, and frost-proof freezer systems for refrigerators in 1958.[30] The 1960s saw a huge variety of new features for refrigerators: automatic icemakers, icemakers inside refrigerator doors, adjustable shelves (in the doors and in the cabinet), vegetable crispers, meat keepers, covered butter and cheese compartments, charcoal filters to soak up food odors, just to name a few.[31] Despite these innovations, the basic cycle of compression, condensation, and expansion remained the same. While each new feature undoubtedly had its fans, non-essential changes sometimes adversely affected the quality of refrigeration that the unit provided and always made them harder to service.[32]

With few compelling innovations, increased size became a crucial selling point for new models of refrigerators, as more storage space had an indisputable effect on people's lives. Between 1930 and 1950, just as people who bought the earliest models began to look for replacement models, the average American refrigerator doubled in volume, even as the amount of floor space they took up shrank.[33] Luckily for the industry, this push to sell Americans large refrigerators coincided with a period when people had lots of extra money to spend. Need more storage space? Buy a new refrigerator. Want to visit the market less? Buy a new refrigerator, and make it a big one. "No wonder young housekeepers who start out with a shiny de-luxe refrigerator are the envy of the neighborhood," noted *Good Housekeeping* in 1950. "They have worlds of extra cold space and food storage convenience. They never have to wonder where to put the eggs or where to store perishables to keep them fresh and appetizing."[34]

Thanks to the length of the postwar prosperity, the growth in the size of household refrigerators continued throughout the post–World War II period and even beyond it. In 1934, a popular guide for the purchase and operation of household appliances recommended one to two cubic feet of refrigerator space for each person in the family.[35] In 1938, the largest refrigerator on the American market held 8.6 cubic feet of food. Most models had capacities between 4 and 7 cubic feet.[36] Over time, refrigerator manufacturers managed to fit more room for food into smaller spaces. By 1947, 88 percent of new refrigerators had 7 cubic feet of space or more.[37] Despite the energy crisis, the largest Frigidaire available in 1972 had a capacity of 22 cubic feet.[38] Lowe's 2007 refrigerator buying guide suggests that American refrigerators have only gotten bigger since then. "Today's refrigerators typically have capacities of 18 to 26 cubic feet," the retail chain informed potential purchasers.[39]

The buying habits of American consumers certainly explain some of this increase in size. Advertising did much to drive this trend. "Here's your new Frigidaire," went the tagline for that company's 1940 ad campaign. "It's big, It's beautiful, It's a bargain."[40] A 1945 Westinghouse ad echoed a theme first heard during the 1920s: "When unexpected guests make a mealtime call . . . you can have a big, well-planned Westinghouse Refrigerator, with special places for everything . . . and a generous section for frozen foods."[41] Women's magazines offered the same pitch for increased size. "An 8-cu.-ft. box will fit in the space formerly required of a 6-cu.-ft. box," explained *Good Housekeeping* in 1949. "The difference in operating cost between an 8- and 10-ft. box is virtually negligible. So don't skimp on the size of your new refrigerator."[42] After all, someday you might need it. Size remained a top priority for refrigerator advertising into the 1960s. "You're a smart, practical woman," declared Amana in 1964. "Probably have your own common-sense idea of how a freezer-refrigerator should look. Chances are you want one with big family space."[43]

Today, America has the largest refrigerators in the world. In the interests of combating global warming, the European Union has collected information about the volume of refrigerators worldwide (since larger refrigerators generally require more energy to run than smaller ones). At the time of this writing, this information, shown in table 8.2, is over ten years old, but if anything, those numbers have grown larger in the United States. According to another set of figures, the average refrigerator volume in the United States was 19.6 cubic feet in 1992 and had increased to 28.6 cubic feet by 2002.[44] Twenty percent of U.S. homes have more than one refrigerator, which means that even these large numbers may not reflect every American family's total refrigerating capacity.[45]

Of course, another common household appliance also keeps things cold.

Table 8.2. Size of household refrigerators worldwide, by volume

Country/region	Volume (cubic feet)
North America	17.50
Canada	16.35
Australia	11.86
Japan	9.80–12.00
Western Europe	9.71
Russia	8.12

Source: Organisation for Economic Co-operation and Development (OECD), *Energy Efficiency Standards for Traded Products*, Annex 1, Working Paper no. 5 (Paris: OECD, 1998), 27.

Note: Figures from countries that use the metric system have been converted from cubic liters to cubic feet, and the results have been rounded.

Many people in the United States own stand-alone freezers, an appliance that began as an appendage to the household refrigerator but eventually struck out on its own.

America Warms to the Freezer

As every shopper discovers, perishable food can spoil even when refrigerated if it is kept too long. The optimal way to keep food waste at an absolute minimum is to keep it frozen. Americans and American refrigerator makers soon decided that they needed large freezers and frozen food to go in them. "Yes, the new 1952 Ben-Hur Freezer offers beauty you're proud to own," begins the text of an ad for that product. "But it's the roomy interior that brings you the most pleasure . . . plenty of room for hundreds of pounds of fresh frozen foods. . . . A 'supermarket' at your finger-tips."[46] Like the ads for refrigerators before them, advertisements for home freezers stressed convenience and choice. "Fresh-frozen," a term that seems meaningless to modern ears, often appeared in the pitch for freezers. This kind of terminology helped prevent the less than perfect taste of frozen food from holding back freezer sales. Instead, manufacturers stressed this particular technology's many other advantages.

Freezers, like larger refrigerators, served as another opportunity for the refrigerator industry to make money from an American public that had already taken to refrigeration. If you willingly sacrificed a little freshness in order to keep things in a home refrigerator, sacrificing more freshness to keep things in a freezer changed things only marginally. Keeping additional perishable food in the home required more cold capacity. As the household refrigerator market became saturated, appliance manufacturers became increasingly interested in selling Americans additional freezing capacity. The small freezer compartments in existing refrigerators became inadequate. Instead, the industry pitched new stand-alone freezers. "The Electric Home Freezer frees the homemaker from the daily grind of three-times-a-day food preparation, and frequent shopping," wrote the National Electrical Manufacturers Association in 1948. "It enables her to put meals on a modern, production-line basis." The association recommended preparing meals in advance, then dropping them in an oven, pot, or broiler shortly before mealtime. Unlike other organizations touting home freezing, the group recognized that freezing might affect the taste of food: "There are some people who believe that freezing does 'tenderize' certain types and cuts of meat, but there is little scientific evidence for this belief." Prepackaged frozen foods, still a relative novelty at that time, did not earn a mention in the entire pamphlet.[47]

Freezing utilized the same basic technology that made mechanical refrigeration possible. The first experiments with frozen food, dating back to the 1860s, used ice and salt to generate cold.[48] Ice making obviously required freezing water, but the freezing of food took longer to develop because keeping food in good condition required lots of learning through trial and error. Ice cream freezers, a common appliance during the late nineteenth century, depended upon ice as the source of cold. Freezing food became a viable business model only when households had enough freezer capacity to store it. "Your Frigidaire Food Freezer does for long-term storage of food exactly the same thing that your Frigidaire Refrigerator does for short-term storage of food," explained that company in 1949.[49] The introduction of significant freezer space in households meant that housewives could not only freeze and reuse their own leftovers but also freeze fresh foods for later use.[50] With a few supplies like freezer bags, freezer wrap, and a marker, homemakers could begin preserving food through freezing by following any number of instructions offered by the freezer company, by the company that sold freezer supplies, or in major women's magazines. When freezing became commonplace, the need for canning, pickling, and just about every other traditional method for food preservation completely disappeared (at least in the sense of being a household necessity). After making ice cubes, freezing leftovers constituted the primary reason to have any home freezing capacity at all, as it took many years for the frozen food industry to perfect the production, transport, and safe preservation of its products.

Clarence Birdseye developed most of the technology that made the modern frozen food industry possible, although frozen foods existed long before Birdseye came along. They did not catch on because early producers froze low-quality food and because the technology did not work well.[51] Birdseye first became interested in freezing foods while living in Labrador during the 1910s. There, he recognized that smaller ice crystals form on fish that are frozen quickly than on those frozen more slowly. More importantly, he recognized that those smaller crystals preserved the food's appearance and taste rather than harmed it.[52]

Of his many innovations, Birdseye's most important frozen food patent, for a multiplate freezing machine, came in 1927. Multiplate freezing made it possible to freeze lots of food quickly, using refrigerant-filled plates that did not come in contact with the food itself.[53] (In this way, it resembled the technology of cold storage, which also separated food from refrigerant in order to prevent tainting.) General Foods bought up Birdseye's company and, more importantly, its patents in 1929.[54] In its early years, General Foods concentrated on introducing packaged frozen food in urban areas, rolling it out in new markets only very slowly. Since consumers had little freezer space available, institutions like cafeterias bought most of the available product.[55]

Besides refrigerated display cases in grocery stores, success for the frozen food industry depended upon larger home freezers. Frozen food first appeared regularly on America's store shelves before most Americans even had enough freezer space to stock significant amounts.[56] Because that household infrastructure did not exist, producers marketed frozen foods much the same way that the first produce shipped long distances had been marketed. In 1932, the *New York Times* called frozen foods a "scientific miracle in home management" and noted the novelty of being able to eat foods like peaches and strawberries out of season.[57] This argument did not convince enough consumers to buy frozen food to keep the industry growing during the Depression. The technological innovations by Birdseye and others made the frozen foods business viable, but the industry did not find a mass market until refrigerators and especially the freezers included in them started to expand in size.

Because of food shortages during World War II, many people overcame their misgivings and tried frozen foods for the first time.[58] Some consumers did not get a good first impression. Failures along the early cold chain for frozen food meant that spoiled food often appeared in grocery stores. The failure of growers to freeze the food that would keep best created another problem. Even apart from this shaky infrastructure, most people could not afford frozen food. Many smaller frozen food producers who lacked the technical knowledge to freeze food and distribute it well went out of business during the late 1940s. The companies that remained operated on a mass scale. The frozen food industry did not turn a profit until the 1950s, when better infrastructure largely solved its distribution problems. This improved distribution also lowered prices, even though by then many consumers willingly paid more for the convenience that frozen food provided.[59]

To become big business, frozen food needed a cold chain entirely of its own. Companies constructed special warehouses to keep the frozen food cold after packaging until it was shipped to the grocery store.[60] A whole new set of railcars with enough insulation to protect frozen food packages on their long journey to market had to be developed. Consumers also had to purchase more freezer space before they could buy enough frozen food to make mass production viable. In the early days of electric household refrigerators, small freezers existed even in the earliest units because the areas closest to the evaporator were always below freezing. The Guardian Refrigerator Company, the forerunner of Frigidaire, had just enough freezing space to hold two ice trays.[61] With the gradual perfection of temperature control over the course of the 1920s and 1930s and the gradual increase in the size of home refrigerators in the United States, it became possible to devote a separate section of the unit entirely to keeping things frozen. This usually required the sacrifice of refrigerator space for freezer space.[62] Looking for extra freezer space, families first turned to locker plants.

Locker plants were individual frozen storage units inside cold storage warehouses or in ice plants with extra space and refrigerating capacity. (As the ice business declined, there were plenty of facilities that met that description.) Locker plants operated the same way that safe deposit boxes do in banks. People came into the plant, turned the key to their locker, and either put in or took out their perishable valuables. In 1941, renting a locker might cost $10 to $12 per month.[63] While there had been some experiments with rental space in early cold storage plants, this industry did not really blossom until the mid-1930s, when the first cold storage facilities consisting entirely of lockers were built in Iowa for the benefit of local ranchers.[64] The industry grew from approximately 250 plants in 1936 to over 5,000 at the beginning of 1944. Much of that growth came through expanding the customer base. Often, hunters rented these spaces to store excess meat from a particularly large kill, but by the post–World War

II period, individual families were renting these units to store fresh fruits and vegetables. This made it possible for urban dwellers to buy perishable food in bulk to take advantage of low prices.[65] On the negative side, people found having to go to a cold storage locker to retrieve food extremely inconvenient. In 1947, the National Food Locker Association estimated that people who rented space made one and a half trips to their lockers each week.[66] The demand for home freezers followed logically as a way to eliminate the need to make any trips at all.

Frigidaire introduced the first household refrigerator with an expanded freezer in 1930. It failed to sell in sufficient numbers, and the company removed it from the market after only a year. By the end of the 1930s, though, all major producers began to offer larger freezing spaces. Creating extra freezing capacity became increasingly easier as the size of the overall refrigerator grew.[67] In a pattern that went back in the days of iceboxes, manufacturers usually placed freezers at the top of their units. (This generally remains true today.)[68] By the late 1940s, manufacturers were pushing separate home freezers so that homemakers would always have enough storage space for frozen foods. Initial sales of Frigidaire's early home freezers proved disappointing. Many customers did not have the space for a separate appliance and continued to rely on the freezers in their refrigerators.[69]

This situation began to change during the early 1950s. In 1952, the industry sold over a million stand-alone home freezers. The size of these freezers grew along with total sales. The seven- or eight-cubic-foot-capacity models of the late 1940s gave way to models with capacities of twelve to fourteen cubic feet in the early 1950s.[70] In 1953, Americans owned over three million home freezers. "Within the last few years," wrote three food technology specialists in a 1953 text, "more and more people in all sections of the country, from all walks of life and at all income levels, have been demanding and getting freezing facilities for their homes. A refrigerator without a separate freezing space and unit is now definitely *passé*. Refrigeration experts now see the market for refrigerators stabilizing while the market for home freezers is on a wide upswing."[71]

But even when Americans had freezers, they did not necessarily want them in order to store prepackaged frozen food. The episode of the classic TV sitcom *I Love Lucy* that first aired on 28 April 1952 illustrates this trend well. Lucy tries to convince Ricky to buy her a home freezer by showing him an advertisement suggesting that the freezer "paid for itself" because it would allow them to buy their meat wholesale.[72] The emphasis on meat storage as a motivation to buy a freezer demonstrates how few Americans bought frozen foods at that time.

By the end of that decade, this situation had completely changed. The American Motors stand-alone freezer from 1959 included trays in the door designed specifically to fit frozen food packages of all kinds, as well as a basket at the bottom to hold large TV dinner boxes. A juice dispenser designed to fit cans of frozen concentrate came with some models.[73] In 1963, the *Freezer Handbook* published by the Good Reading Rack Service recommended: "It's always good household management to keep an ample supply of the commercial frozen foods you use most often. You'll have latitude in menu planning, [and] save yourself a hurried trip to the grocery when that extra guest drops in for dinner."[74] Despite this advice, most of that publication explained how to freeze your own food rather than how to make best use of the prepackaged kind. A guide to a Sears Coldspot refrigerator from the late 1960s begins to suggest the logic by which Americans live today. "In many cases," it argues, "pre-frozen foods purchased at your favorite store are actually 'fresher' than those that you might buy to freeze at home," since flash freezing happened at peak freshness. Home freezing, in contrast, depended upon the condition of the foods when frozen. The pictures inside the Coldspot booklet illustrate nothing but frozen food packages and store-bought meat still in its shrink wrap.[75]

Thanks to these two kinds of food, prepackaged and home-frozen, Americans owned thirteen to fourteen million stand-alone home freezers in 1965.[76] The increased acceptance of frozen foods at home made them increasingly acceptable outside the home too. By the mid-1960s increased demand from institutional users like restaurants and cafeterias accounted for most of the increase in the output of the frozen food industry. The first frozen meals from precooked recipes also appeared around this time. (Instead of a frozen chicken patty, you might now buy a chicken pot pie.) This development both made chefs less skilled and standardized portions, with frozen foods packaged in predetermined serving sizes.[77] Today, even in a mid-range family dining establishment like Applebee's, just about everything served is precooked and frozen.[78] Frozen-food-industry representatives claimed that frozen food tasted better than fresh food since it was preserved right after it was picked. True or not, this technology changed the taste of the food that most Americans ate every day.

The growth of frozen foods and the freezers that they required also had more mundane ramifications for American homemakers, even before free-standing freezers. Frost accumulates on the cooling elements of a refrigerator as water in the air passes over the evaporator or on the colder parts of the inside that the evaporator chills below freezing. This led to the rather tiresome and inconvenient chore of having to defrost the refrigerator. When done right, this required shutting the entire appliance down until the frost melted (at least partially) and the icy buildup could be safely removed. In 1927, with the introduction of the Monitor Top, General Electric recommended defrosting overnight. The ice melting on the inside provided enough cold to keep the food inside from spoiling during this process.[79] Unfortunately, impatient consumers often tried to remove the ice built up on the freezer with ice picks or other sharp tools and often wound up puncturing the evaporator coil, releasing all of the sometimes noxious refrigerant and ruining the refrigerator. By the late 1940s, expensive household refrigerators could defrost themselves by shutting down automatically.[80] In the early 1950s, refrigerator makers began to introduce automatic defrosting devices as extra equipment. These devices were timers that shut down refrigerators overnight and turned them back on automatically. Once these devices became available, outside sellers began advertising them for older refrigerator models, since nobody liked defrosting a refrigerator.[81]

Refrigerator makers did their best to solve the defrosting problem from early on in the history of household refrigeration. With the most advanced early models, owners could begin the defrosting process simply by flipping a switch, but this was still inconvenient for many users. The Hotpoint Refrigerator-Freezer borrowed from the drip pans of iceboxes. "The General Food Storage Compartment does not require defrosting as it does not accumulate frost," explained a catalog from 1950. "Instead, the tiny droplets of moisture formed on the inner walls will be collected in the drip tray underneath the compartment. The small amount of water that collects will be evaporated by the heat of the refrigerator mechanism."[82] The freezer and the refrigerator shelves still required defrosting, though. The 1953 Westinghouse frost-free refrigerators were fully automatic. A heater melted the frost around the freezer, which then dripped down and evaporated without the user even needing to clean a tray.[83] The refrigerator still accumulated some frost, however, affecting the efficiency of the appliance. Finally, in 1958 Frigidaire introduced its first Frost Proof refrigerators, in which no frost ever accumulated. These models defrosted automatically, without shutting the refrigerator down, by means of electric heaters alongside the cooling coils where frost builds up.[84] Although the Frost Proof was originally intended as a top-of-the-line model, over 100,000 of them sold in one year.[85] Today, all refrigerators with the exception of very small models meant for cramped apartments or extra cold drink capacity are frost-free. As all the burdens associated with owning a refrigerator disappeared, these appliances became more popular worldwide.

Refrigerators and Freezers around the World

No other country took to the household refrigerator as fast as the United States. By 1941, just twenty-five years after its introduction, 45 percent of American households owned a mechanical refrigerator.[86] Italians, on the other hand, insisted (and still do to a great extent) on shopping daily for their food at local markets. As late as 1957, only 11 percent of Italian families owned household refrigerators.[87] That same year, the *Times* of London reported on refrigerator adoption rates around the world; in Great Britain, fewer than 10 percent of families had a refrigerator (table 8.3). In 1965, only 38.2 percent of English families had refrigerators, and the ones they had were much smaller than American models because they had to fit into smaller living spaces.[88] This delay in adopting home refrigeration reflects not only the hesitance of Europeans to buy a technology that many felt they did not need; it reflects the slowness of electrification there as well.

Carl Linde of Germany's Munich Polytechnic College had pioneered mechanical refrigeration during the late nineteenth century, making it commercially viable for the first time and improving the technology over his entire career. Despite this head start, Germany lagged significantly behind the United States when it came to the production and adoption of the household refrigerator in the middle of the twentieth century. As late as 1951, the West German Parliament considered taxing refrigerators as a luxury good. By the early 1960s, however, almost 54 percent of West German households owned refrigerators. While prosperity and the increased popularity of frozen foods had much to do with this change, it also helped that Linde's old firm began producing refrigerators at a new factory in Kostheim in 1956. By 1959, the firm made models for export. Refrigerator ownership in Europe in the late 1950s and early 1960s resembled that of the United States in the late 1920s and early 1930s. They became very popular in Europe right after the *Times* of London published the figures shown in table 8.3.[89] In France, only 3 percent of homes had refrigerators in 1950. That figure had reached 90 percent by 1975.[90]

Unique aspects of American economic life helped make European nations look slow in adapting refrigeration by comparison. The United States took to

Table 8.3. Refrigerator adoption rates around the world, 1957

Country	% of homes with refrigerator
United States	90+
Canada	84
Australia	70
Sweden	50
New Zealand	26
Denmark	25
West Germany	14
France	12
Italy	11
Great Britain	<10

Source: *Times of London*, 10 Dec. 1957, 8.

refrigeration faster than anywhere else in the world because of its wealth, its abundant and available amounts of perishable food, and the economic incentives provided by its relatively unencumbered free enterprise system. This gap in refrigerator adoption rates also explains a lot about the countries that did not take to this technology so quickly. First, Europe's general historical resistance to refrigeration developments of all kinds suggests cultures that preferred fresh foods. On a technical level, Europeans manufacturers did not have the economies of scale needed to make inexpensive refrigerators for the working class. Workmen at a Frigidaire plant in the United States could buy one of the models coming off their assembly line for a month's wages. An identical refrigerator in France cost the average workman eight or nine months' wages.[91] (And it was not just in Europe that refrigeration was more expensive; in Chile in 1960, an average refrigerator cost an affluent white-collar worker ten weeks' salary.)[92] Similarly, Europeans had little reason to buy freezers because frozen food production grew so slowly there. In 1960, 13,000 tons of food were quick-frozen in France, as compared with three million tons in the United States. France lacked the infrastructure to turn frozen food into big business.[93]

Despite such problems, the rest of the world began to develop an interest in refrigerators long before the gap in refrigerator adoption rates started to close. In 1939, the United States produced 90 percent of the world's household refrigerators. In 1958, the United States produced only 37 percent of the world's refrigerators.[94] By the early 1960s, the total number of refrigerators constructed in Europe actually exceeded the number produced in the United States. The explosion in home refrigerator use, not only in Europe but also elsewhere, explains why even as refrigerators have become increasingly efficient, the total amount of energy expended upon refrigeration worldwide has increased significantly. The increased productivity of refrigeration has made it so cheap that for the first time countries besides the United States have begun to utilize this technology the way Americans always have.[95]

This does not mean that most countries now have refrigerators as big as the ones in the United States. European refrigerators in the 1950s reflected that continent's social and economic realities. The largest refrigerator on the U.K. market in 1960 was the same size as the smallest one available in the United States. European refrigerators were also smaller in order to fit in smaller living spaces, and cheaper.[96] As European wealth grew and housewives increasingly took on outside jobs, the size of the European refrigerator grew, just not as far as American models.[97] European refrigerators are still smaller than those in the United States, but the size of refrigerators on both continents expanded at similar rates in order to store the goods that an increasingly efficient cold chain can supply for them.

Today, household refrigerator capacity in Asia has also begun to rise. "People are moving into newly renovated apartments, so they want a pretty new fridge," explained a clerk in a Shanghai appliance store to the environmental journalist Bill McKibben in 2011. "People had a two-door one, and now they want a three door."[98] Refrigerators, in other words, have become an object of conspicuous consumption there, just as they were in America eighty years earlier. Besides growing in size, the market for refrigeration in China also has an enormous potential to grow. As of 2005, only 12 percent of rural Chinese families owned refrigerators, as compared with an 80 percent ownership in urban areas. This leaves plenty of room to expand the reach of refrigerators even in China's developed areas.[99] As more people in India enter the middle class, experts have begun to question whether the power grid there can serve all the new television sets, computers, and refrigerators that these up-and-comers will want to purchase.[100]

New refrigerator buyers in Asia have purchased lots of unnecessary refrigerating capacity, just as Americans have done for years. In South Korea, for example, consumers can buy special refrigerators just for kimchi so as to keep their favorite condiment, made from pickled cabbage, cold. Ironically, as the *Korea Times* reported in 2011, these appliances "aim at mimicking the conditions of a clay pot buried in the ground." Koreans created kimchi as a means of food preservation, but South Koreans believe that cold kimchi tastes better than kimchi at room temperature because the flavor changes less over time. For this reason, the kimchi refrigerator has become the most wanted household appliance in Korea despite the fact that it preserves a food product that does not require refrigeration.[101]

Taste, variety, and convenience have driven the development of the ice and refrigeration industries from its earliest days. The completion of the modern cold chain has made all these advantages available to more people around the world than ever before. The path from the early days of the natural ice industry to refrigerated shipping containers and airplanes presented the industry with many technological obstacles. Most of the technologies created to solve these problems were adopted only gradually. Despite the variety and slowness of these changes, the results of these developments have changed the world. The cold chain comes with costs, but those costs pale in comparison with the many benefits—more food, a greater variety of food, fresher food—that these technologies have brought to the countries that make extensive use of it. Because of its usefulness, refrigeration has become part of their respective cultures, and in no place more so than America.

CONCLUSION

Refrigeration, Capitalism, and the Environment

When Hurricane Katrina struck New Orleans in 2005, the resulting loss of electricity meant that almost all refrigerators in the city suddenly became useless, regardless of how much damage a given neighborhood suffered. Since few people had emptied their kitchens before evacuating, most came home to find putrid, maggot-infested food in their kitchens. Their refrigerators were too filthy to restore to regular use, so the people of New Orleans, rich and poor alike, sealed them with duct tape to prevent the doors from swinging open and then pushed their old refrigerators out onto the curb, assuming someone would eventually come to take them away.

In January 2006, New Orleans poet Andrei Codrescu introduced the wider world to the discarded refrigerators of Hurricane Katrina in the pages of the *New York Times*.[1] Codrescu did not particularly care about the death of so many appliances at once. He cared more about the messages that the people of New Orleans had painted upon them: "'Chem Trails Are Real: Weather Control Is Here,' was scrawled on a refrigerator below a drawing of a jet leaving behind what looked like a trail of poison. Another fridge warned severely: 'Do Not Open: Cheney Inside.' Inside others one could find President Bush, Condoleezza Rice, Ray Nagin and Michael Brown doing obscene things with the maggots and with each other. In a short time, there were thousands of art works in the city, an exhibition that stretched for miles, that had no official opening, that was constantly in progress."[2]

You can still see these messages in the photos accompanying that *Times* article or on Flickr sets on the Internet, but the canvasses upon which these messages appeared deserve as much attention. Thanks to the incredible cornucopia of perishable food at affordable prices that the cold chain provides, even the poorest households in the United States own refrigerators large enough to become impromptu street-side billboards. Such widespread and extensive use of refrigeration came as a result of many technological developments that occurred around the world and cultural developments mostly peculiar to America.

Even without spray-painted messages upon them, people can still "read" household refrigerators for signs of what they reveal about American society. For a 1985 article, the anthropologists Bruce Hackett and Loren Lutzenhiser showed photos of the insides of refrigerators to people other than the owners of those appliances. Then they asked the interviewees what the contents of those refrigerators told them about their owners. "The refrigerator can be likened, sometimes *is* explicitly likened to a piece of kinetic sculpture," they wrote, "a household art-object-in-process, constantly modified (or *not* modified) so that it looks, feels, *is* right. It is true that we do not gather audiences to gaze upon 'the fridge,' or gather refrigerators together in galleries. . . . But then it is not really clear *why* refrigerators are not the stuff of galleries; in fact it isn't really clear that they aren't."[3]

Katheryn Krotzer Laborde, writing about the discarded refrigerators of Hurricane Katrina, also focused on the messages sent by what we keep inside these appliances. "Are they filled with fresh vegetables?" she asks. "Medication? Uncovered plates of half-eaten food? Leftovers stored in neat, matching, perhaps even labeled containers sidled next to packaged chicken, thawed and ready for the next night's dinner? Or is there nothing but the remnants of take-out, one half meal piled on top of the other in Styrofoam go-boxes?"[4] The contents of our refrigerators reveal how much disposable income people have and how they prioritize their time. The size of their refrigerators reveals their changing relationship to the natural world. Because both these aspects of our refrigerators reflect the cold chain that makes provisioning them possible, they also reflect how Americans eat and who they are.

With respect to how people eat, refrigeration allows consumers to take full advantage of the modern mass production of foodstuffs of all kinds. Before food even reaches our homes, food producers provide refrigeration to make sure that their products will be marketable at the end point of the cold chain. Without refrigeration, perishable food could never have become so plentiful in the modern world. Theoretically, there is no limit to the number of manufactured objects that any consumer can own. But the amount of food that a consumer buys is limited by the size of an individual's stomach. Even the heaviest among us can only eat so much in a single day. Food companies want to convince consumers to spend more than just the cost of the minimum number of calories needed to sustain life. Food preservation, especially through refrigeration, encourages consumers to spend more on food since it increases variety at all times of the year. After all, the calories people consume to provide energy do not depend upon geography or season. People in industrialized countries value the opportunity that refrigeration offers and pay the extra price for this variety. They have come to expect any fruit or vegetable to be available at the grocer down the street even during the winter. As the American beneficiaries of refrigeration have done so for many years now, they generally take these advantages for granted.

The incredible variety of food choices that the cold chain makes possible allows people to express their individuality at the table. "We can say 'You are what you eat' only when we have some minimal choice about what we're eating," the writer Adam Gopnik observes.[5] In fact, Americans now have more choice in their diet than has ever been possible in the history of the world. The addition of more perishable food to diets in the industrialized world has made that wide selection possible. Eat the same thing every day, and eating becomes humdrum. Eat a great range of foods, and eating becomes one of the highest forms of pleasure available to human beings. In the United States, which takes ingredients and recipes from around the world and creates fusion cuisines that spread back across the globe, the diversity of available food can leave people from other cultures breathless.

Most other developed countries do not have the same degree of food variety available because they have not embraced refrigeration like the United States has. Lesser-developed countries do not have the infrastructure or a wealthy enough population to make extensive use of this technology. In many well-developed countries, the failure to utilize the cold chain to as great an extent as Americans have stems from differences in culture. This difference in attitude towards refrigeration between the United States and the rest of the world existed even during the very early history of the cold chain. As previously mentioned, Europeans have never used as much ice in drinks as Americans do. The *Shenandoah (Virginia) Herald* told its readers in 1879, "The Englishman and the German fairly shun ice, though placed in easy reach of boundless quantities of it, and a Frenchman would as soon think of taking an emetic as of chilling his stomach with an iced draught."[6] For many years, only Americans ever dropped ice in their drinks. "I think that there is but a single specialty with us," Mark Twain wrote of his country's taste in liquid refreshment, "only one thing that can be called by the wide name 'American.' That is the national devotion to ice-water. . . . I suppose we do stand alone in having a drink that nobody likes but ourselves." Twain complained bitterly about the poor-quality ice water he received while traveling in Europe.[7] Finding any ice water throughout the

Continent must have been difficult, since few Europeans drank it during the late nineteenth century except in the kind of hotels that Americans like Twain frequented. Twain countered the European argument that ice water impairs digestion by quipping, "How do they know?—They never drink any."[8]

Russia once resembled the United States in its love of cold things. "The Russians have accustomed themselves to use a prodigious quantity of ice for domestic purposes," a German observer wrote in 1842. "They are fond of cooling all their beverages with ice; indulging themselves freely in the frozen juices which are sold all the summer in the streets of all their towns; and drink not only ice-water, ice-wine, ice-beer, but even ice-tea."[9] While Russia had easy access to natural ice, it did not develop a modern cold chain maintained by mechanical refrigeration until very recently. A foreign consultant for the American Society of Heating, Refrigerating and Air-Conditioning Engineers reported in 1965 that "the output of domestic refrigerators does not satisfy demands" in the Soviet Union.[10] The most common household refrigerator in the USSR did not even come with a freezer until 1960. As a result, Russians could only buy ice cream, another manifestation of their love of cold foods, in individual servings from pushcarts refrigerated by dry ice.[11] This situation suggests the importance of infrastructure for making any cultural predilection for the consumption of cold things possible. That infrastructure came earlier to capitalist countries than communist ones because capitalist economies foster markets that respond to the demands of consumers in search of a better life, and refrigeration often plays an important part in making people's lives better by improving their diets.

"Roast Beef and Apple Pie"

The history of the cold chain includes both the development of many new technologies and the increasing availability of different types of food. The growth of the cold chain depended upon technological innovations all the way from the point of production to the point of consumption. Unlike other industries where new inventions sometimes create sudden changes that transform societies overnight, almost all the technological changes associated with the creation of the cold chain came gradually. These gradual transformations, however, sometimes brought very sudden changes in the eating habits of people in America and around the world. High-quality meat at low prices, strawberries out of season, cold drinks replacing warm ones—these effects explain refrigeration's obvious historical significance. America's ability to control the cold, and the foodstuffs which cold technology brought to people's tables, made it possible to feed the growing population of a rapidly industrializing country.

As cold chains have spread over the planet, they have helped feed the rapidly growing population of the world. Obviously, refrigeration has not put an end to starvation. Instead, some countries have too much food, while others do not have enough. Although this is an unfortunate situation, the pioneering early-nineteenth-century demographer Thomas Malthus imagined a much worse future for the world. To be fair to Malthus, his famous apocalyptic predictions about limits to future population growth contained two possible outs for marginal countries: either they had to produce more food or they had to acquire it from somewhere else.[12] In the late nineteenth century, many farmers across the world began to industrialize their operations so that they could produce more and meet the emerging global demand. Efficient, intensive mechanized agriculture would have been useless if not for the ability to transport the surpluses it created to the markets that needed the extra supply. Refrigerated transport, like the transport of food in general, allowed countries that could not produce their own perishable foods to acquire it from abroad.

While the diets of poor people improved with the introduction of refrigeration, some perishable foods remained out of their reach because increased variety made it possible for producers to charge more—sometimes much more—for high-demand or hard-to-acquire perishables. But even with occasional price inflation, some foods that had once seemed like luxuries (oranges, for example) did become affordable for the typical American. Working-class people with little disposable income for food ate better and lived longer because refrigeration increased the supply of available perishable goods, which in turn kept prices down. Once the prices of perishable foods were low enough that nearly anyone could afford them, every American could seemingly live like a king. The German sociologist Werner Sombart captured this reality in his classic *Why Is There No Socialism in the United States?* with his remark, "All Socialist utopias came to nothing on roast beef and apple pie."[13] Of course, the supply chain for both apples and beef depend upon refrigeration, at least on a mass-produced scale. Sombart intended this famous line as an exaggeration; even he stressed that material abundance constituted only part of the answer to the question he raised in the title of his book.

Ample perishable food became a symbol of American plentitude even before the cold chain fully developed. As the cold chain developed and improved, food producers replenished that cornucopia faster than anyone in the mid-nineteenth century ever thought possible. Like the original cornucopia of Greek myth, the wide availability of this bounty at first seemed like divine intervention. Then this kind of bounty became commonplace. Roast beef and apple pie—affordable and common in the United States, even in Sombart's time—have never served as symbols of wealth in the United States. They better represent the wealth it took to create the effective cold chain that keeps America's natural bounty fresh. That food, in turn, literally made every other part of the capitalist system function, since it provided calories to workers who produced goods of all kinds.[14]

This helps explain why the eradication of famine often accompanies the integration of markets. The more sources that a country has for food, the less likely it becomes that that country will ever run out of food.[15] Despite substantial progress towards the integration of world food markets since 1917, serious famines have devastated many communist countries. While many of these disasters came about as a result of low food production, the ability to preserve perishable food effectively might have mitigated the disasters that Stalin's Five-Year Plans, China's Great Leap Forward, or North Korea's isolationist policies ultimately caused. China today of course has become more capitalistic than many capitalist countries, and its interest in refrigeration reflects that fact. Industrialized production has made it possible for more Chinese to buy refrigerators. The longest and most efficient cold chains in human history have made it possible for the Chinese to stock ever-larger refrigerators with perishable goods from across the globe. With globalization, every country that can afford to pay can now eat like Americans do, consuming a diet of foods produced elsewhere and preserved by refrigeration on their journeys from all over the world.

Second Nature

Over the course of his legendary career, the business historian Alfred D. Chandler, Jr., outlined the organizational changes within growing companies, both in the United States and in other early industrializing countries, that helped them become large.[16] According to Chandler's thesis, changes in administration were necessary to implement the improvements that made prolonged growth possible. Many firms embraced economies of scale, which allowed them to produce goods more efficiently as they grew larger. Other firms produced a wide range of products from a single resource. This is known as economies of scope. A few of the many firms that made the cold chain possible employed strategies similar to these. In contrast, most of the firms associated with refrigeration technology stayed fairly small or had to diversify into related but separate industries (like air conditioning) in order to become significantly larger. No firm could gain control over the entire cold chain because a wide variety of technologies were needed at different places between the point of production and the point of consumption. No firm could achieve the efficiency or the funds needed to make complete vertical integration possible.

The variety of enterprises that constituted America's industrial economy from 1870 to 1930—a high-growth period for both the refrigeration industry and industry in general—explains the growth of the American ice and refrigeration industries better than Chandler's model. All the manufacturers in the economy at that time can be classified into one of four groups: custom, batch, bulk, and mass production.[17] Examples of all these strategies existed during the development of the cold chain. Custom production—the one-off, on-demand creation of large machines—describes the development of large-scale ice machines perfectly. It also explains the way packers built early refrigerated railroad cars. Batch production—which creates lots made of various sizes from which customers can choose what they want—describes the way that refrigerating machinery makers built and marketed the smaller ice machines of a slightly later era. This label also fits the production of tools designed to aid the natural ice harvesting industry during the late nineteenth and early twentieth centuries. Bulk manufacturing, the mass production of basic necessities, best describes the production of ice itself. After all, ice was both a commodity and a food in the early years of the industry. It also describes the manufacture of iceboxes, which were built in factories once they were no longer made of wood. The last category, mass production, has attracted the most attention from historians. Of all the possible examples related to the ice and refrigeration industries, the production of electric household refrigerators starting in the mid-1920s best represents this category.

Each link in the cold chain adapted available technologies to meet the commercial, cultural, and geographic demands it faced in order to bring perishable products from the point of production to the point of consumption. The ascending stages of cold chain creation—creating the cold, managing the cold, controlling temperature with precision, and the expansion and extension of volume and reach—came when technology made it possible to overcome the primary natural obstacles in making these chains efficient and effective. The leaders of the ice and refrigeration industries did not make a conscious effort to create the best distribution system available for perishable food. Under capitalism, as entrepreneurs pursued their economic self-interests, it just worked out that way. At the same time, consumers pursued the goals of affordability, convenience, and satisfying their palates. While this abstract dynamic might apply anywhere, only the United States had the technological know-how, natural resources, and cultural motivation to become the beginning, middle, and end of so many food chains that run throughout the world.

Food chains in the historical or sociological sense are linear: "from farmyard to shopping cart" is the subtitle of a recent collection on the history of this subject.[18] Food chains in the biological sense are cyclical. These webs could not exist without the transfer of energy from creature to creature or the transfer of energy between the sun and the earth. Historical or sociological food chains are also unidirectional, which means that consumers do not return the food they consume to producers, at least not directly. The money, however, travels in cycles. Companies that create elaborate food chains, especially cold chains, benefit when workers have the resources to pick what they want from the cornucopia that those firms can provide. The money used to pay for that food then circulates throughout the economy, like the energy in a biological food chain.

The inescapable limitations imposed by the laws of nature further separate the ice and refrigeration industries from so many of the other industries that Chandler described. Western and environmental historians sometimes refer to the idea of "second nature" to describe the improvements that humans have made to urban and rural landscapes.[19] The term refers to the kinds of infrastructure that industry created to make use of the abundant natural resources that it wanted to exploit. Second nature often involved using one set of resources to get at another. For example, the growth of the dressed meat industry in Chicago depended upon the natural ice industry. Otherwise, the surplus of meat butchered there never would have lasted all the way to markets in the east without spoiling. Further technological developments along the cold chain such as cold storage and, eventually, mechanically refrigerated railroad cars essentially completed the conquest of nature that nineteenth-century cold entrepreneurs like Frederic Tudor, Gustavus Swift, and Thomas Mort originally conceived. With the completion of the cold chain, now only the natural lifespan of perishable goods under refrigeration limit the possibilities of second nature.

We have greatly benefited from extending the life of perishable foods, but the same technological developments that have made this possible also created ill effects that those early entrepreneurs never imagined either. Most consumers appreciated the availability of out-of-season produce too much to worry about freshness. Eventually, many forgot how fresh food really tasted. There are also more dramatic adverse effects. In expanding the reach of markets and the demand for perishable foods of all kinds, cold chains have put some species at risk. Would there still be a demand for endangered blue fin tuna, for example, if fishermen could not keep their catch fresh as it travels to market in Tokyo? If the environmental effects of the energy used to produce refrigeration threaten to make life itself increasingly unbearable, do the benefits of this technology still outweigh the costs? Second nature can destroy nature, especially as much of the world's population becomes increasingly industrialized and demands an industrialized diet to match.

Evaluating the Impact of Refrigeration

In 2007, New Yorker Colin Beavan and his family began an experiment. They decided to live for an entire year having no net impact on the environment. This meant creating no trash (so no packaged food), using no product made with toxins (so no laundry detergent), buying nothing new, and engaging in no activity that used a power source that generated carbon dioxide emissions that would promote global warming. They lived for a year with no airplanes, air conditioning, or household refrigerator. This last deprivation proved one of the hardest parts of the experiment, as Beavan explained on his blog a few days after disconnecting the one in his kitchen. "No fridge is hard so far," he wrote. "The milk goes bad in a day. . . . Arugula turns yellow in two."[20] Despite their own ground rules, Beavan's wife Michelle found it necessary to take ice from a neighbor's apartment to keep their child's milk cold in a used ice chest they acquired. "There's a reason people have refrigerators," she explained when they gave up trying to find an alternative to mechanical refrigeration and began to borrow ice.[21]

Unable or unwilling to give up refrigeration entirely, other environmentalists have tried to limit the amount of energy consumed through the refrigeration and transportation of their food by eating more food produced in their local areas. The novelist Barbara Kingsolver and her family are perhaps the best-known examples of people engaged in this activity because of the success of her book about their dietary experiment in local eating, *Animal, Vegetable, Miracle*. Kingsolver's family did not even try to give up their household refrigerator. They took a different approach. Kingsolver's book simply tries to convince people to become locavores (sometimes spelled "locovores" or "localvores"). Locavores commit to increasing their intake of local food. The exact definition of the term "local" depends upon what is available in the immediate area. You do not have to grow your own food in order to live the locavore lifestyle. You just have to live near people who do. When you eat local, your eating habits do less damage to the environment than they might otherwise.

Locavores often condemn the amount of energy consumed for refrigeration. "Americans put almost as much fossil fuel into our refrigerators as our cars," notes Kingsolver's husband, Steven L. Hopp, in a sidebar to her narrative.[22] Refrigerators do use more energy than any other home appliance—approximately 20 percent of an average household's consumption—since their motors must run day and night.[23] Locavores also lament the effects of energy consumption all along the food chain. According to one study, the average piece of supermarket produce in America travels 1,500 miles from picking to point of sale.[24] Unfortunately, focusing exclusively on food miles can mislead people. Bananas, for example, travel across the world to supermarkets in the northern parts of the world. Because they almost always go by boat, they actually have a very small carbon footprint.[25]

Similarly, sacrificing the many positive effects of refrigeration for the sake of the environment does nothing to limit global warming overall. The typical refrigerator consumes 9,000 calories of energy per week, and anything older than newer, energy-efficient models consumes twice that. While that may sound like a lot of energy, a ten-mile round trip to the farmer's market will consume 14,000 calories of fossil fuel energy if you drive.[26] Food production also generally has a much bigger carbon footprint than the way something gets transported.[27] In other words, refrigeration constitutes only one of many aspects of food chains that expend energy or release greenhouse gases that can adversely affect the environment. The popularity of refrigerators even among locavores illustrates the need of society in general to balance a desire for sustainability against the many advantages that refrigeration provides.

A fair balancing of the costs and benefits of refrigeration would have to consider the fact that all the food which refrigeration preserves for consumption makes it unnecessary to produce more food to replace it. People today can hardly conceive of the extent of food waste that existed before the era of refrigeration. In the seventeenth century, a quarter of perishable agricultural products rotted in the fields, and eggs and milk spoiled unless consumed quickly or preserved by traditional methods. Ice, even when available, was not used extensively until the natural ice industry began in the early nineteenth century.[28] In 1917, the U.S. government estimated that 36.25 percent of food was wasted before it could be consumed.[29] That figure had dropped to 27 percent by 2008.[30] The waste that refrigeration prevents means less fossil fuel burned to produce and transport replacement food. Refrigeration also makes it easier to feed people who do not have enough food by increasing the available supply overall.

When poor people have better access to a wider variety of foods, their diets improve. Since refrigeration increased the supply of food, prices dropped too. As a result, the diets of workers with even the lowest wages improved remarkably. One study suggests that after its widespread adoption in 1890, refrigeration increased people's food intake by 5,500 calories and 400 grams of protein per capita per year. This added nutritional benefit, in turn, was partly responsible for an increase in the average height of American adults that happened at the same time.[31] Refrigeration had this effect upon the general population because less spoiled food meant that they had more food available to eat. Preventing food waste can itself improve the environment, since decomposing food in landfills emits greenhouse gases.[32]

Preventing wasted food also prevents the waste of labor needed to grow that food. Historically, this kind of work has proved unrewarding both financially and in terms of job satisfaction. It remains so now. This explains why immigrants (legal or otherwise) dominate agricultural labor in America today. In lesser-developed countries, the cold chain can save people from drudgery in the kitchen that is simply unheard of in the United States. "Still living close to the source of your food, you often don't have a refrigerator or freezer," explains the former chef and travel television host Anthony Bourdain in reference not just to a location in rural Mexico but to similar places around the world. "Equipment and conditions are primitive. You can't be lazy—because no option other than the old way exists. Where there are freezers and refrigerators, laziness follows, the compromises and slow encroachment of convenience. Why spend all day making mole when you can make a jumbo batch and freeze it? Why make salsa every day when it lasts OK in the fridge?"[33] As in the United States, refrigeration gives Mexican women who benefit from this labor savings the chance to do other things to support themselves and their families.

The modern cold chain not only makes it possible to distribute food more efficiently; it helps with efficient food production too. Under the law of comparative advantage developed by the economist David Ricardo, countries can specialize in the production of a few foods (or perhaps no food at all) and still eat well because they can trade their surpluses with other countries. At least in theory, this works to the advantage of both sides. The ability of regions or countries to specialize in particular kinds of agricultural production has also meant that they could produce their food more efficiently. This again leads to more food produced overall. Without this kind of trade, a country with few natural resources like Japan would have to eat a local diet consisting primarily of rice, potatoes, sweet potatoes, buckwheat, and vegetables. It would also face periodic famines and widespread malnutrition.[34] Perhaps more importantly for understanding the economic history of the United States, comparative advantage has meant that different sections of the country could specialize in growing the food best suited for their particular climate. Some parts of the United States specialize in producing food (perishable and otherwise), while other parts of the United States that are not nearly so fecund turn their marginal agricultural land over to other uses entirely. After all, how else could Arizona ever get the fruits and vegetables that its citizens need to live healthy lives?

So rather than condemn the energy used for refrigeration and food transportation outright, environmentally minded people everywhere should work towards mitigating the impact of this incredibly useful technology. These kinds of improvements have already happened in refrigeration history. In 1974, for example, Sherwood Rowland and Mario Molina discovered that the extensive use of CFCs (chlorofluorocarbons, a generic term for Freon and related fluorine-based gases) posed a threat to the earth's ozone layer. Du Pont had developed these gases for use in refrigeration, but the applications for which they were used only expanded over time. The ozone layer prevents most of the sun's UV rays from making it down to the earth's surface, where they can cause skin cancer in humans, among other problems. A British expedition to Antarctica confirmed Rowland and Molina's research in 1985, when they discovered a hole in the ozone layer there.[35] The Montreal Protocol of 1987 banned CFCs and other ozone-depleting substances. Since then, the total amount of such gases in the atmosphere has decreased by 97 percent.[36] In 2010, United Nations scientists declared that depletion of the earth's ozone layer had stopped. As these gases dissipate, the layer will return to its 1980 levels some time between 2045 and 2060.[37]

Similarly, in the 1970s California became the first state to impose energy efficiency standards on refrigerators. Refrigerator manufacturers fought these standards, arguing that they would add too much to the price of their appliances. Once such standards took effect in California, the rest of the country followed suit. Once that happened, refrigerator manufacturers set their engineers loose upon the problem. As a result of technological improvements, the price of the average American refrigerator has decreased by half since then, while its energy use has decreased by two-thirds. Unfortunately, the average size of these units has increased another 10 percent over the same period.[38] That means much of the savings in energy costs have gone to making our refrigerators larger still. If environmental awareness spreads, technological improvements in the future could lead to better results. Decreasing the carbon footprint of refrigeration will become a more desirable public good in its own right as the world becomes more concerned about the effects of global warming.

These success stories should give everyone cause for hope, but the environmental problem with newer refrigerants will be even tougher to solve. The development of other refrigerants that could serve as replacements made the ban on CFCs possible. But those refrigerants had their own problems. Hydrofluorocarbons (HFCs), the refrigerant family that replaced CFCs, are, according to the environmental group Greenpeace, "thousands of times more powerful than carbon dioxide" in producing greenhouse gases and as of 2005 were responsible for 17 percent to the total global impact on climate change.[39] Scientists are currently developing a new generation of natural refrigerants that will not cause global warming, thereby offering the prospect of another transition to new refrigeration technology in the years ahead. These new refrigerants have an added bonus of being cheaper and more energy-efficient than their synthetic counterparts.[40] Most of the impact of the switch will come from commercial refrigeration, further up the cold chain from domestic refrigerators, where after 200 years of refrigeration history there is still much room for improvement. So the evolution of the refrigeration industry continues.

Notes

INTRODUCTION: The Cold Chain

1. Francis Bacon, *Natural History*, in *Works of Francis Bacon* (London, 1826), 1:270.

2. Tom Shachtman, *Absolute Zero and the Conquest of Cold* (New York: Houghton Mifflin, 1999), 22–23.

3. Sue Shepherd, *Pickled, Potted, and Canned: How the Art of Food Preserving Changed the World* (New York: Simon and Schuster, 2006), 30, 66, 190–91.

4. Sidney W. Mintz, *Sweetness and Power: The Place of Sugar in Modern History* (New York: Penguin, 1986), 117, 123.

5. Tom Standage, *An Edible History of Humanity* (New York: Walker, 2009), 159.

6. Of course, any perishable food will taste different as it becomes less fresh. What I mean here is that you cannot taste the preservatives, as you would with salted or pickled food.

7. R. D. Heap, "Cold Chain Performance Issues Now and in the Future," IIR/IRHACE Innovative Equipment and Systems for Comfort and Food Preservation Conference, Auckland, New Zealand, Feb. 2006. I have only seen the term "cold chain" used in refrigeration circles. Nonetheless, I think it has great value for historical analysis too.

8. Shane Hamilton, "Analyzing Commodity Chains: Linkages or Restraints?" in *Food Chains: From Farmyard to Shopping Cart*, ed. Warren Belasco and Roger Horowitz (Philadelphia: University of Pennsylvania Press, 2009), 19.

9. *Milwaukee Journal*, 4 Nov. 1894.

10. Walter R. Sanders, *Selling Ice* (Chicago: Nickerson and Collins, 1922), 15.

11. Simon N. Patten, *The New Basis of Civilization* (New York: Macmillan, 1910), 22–23.

12. For more on this subject, see the essay on sources at the end of the book.

13. Peter Coclanis, "The Burdens of the (Re)Past," *Agricultural History* 72 (Autumn 1998): 669.

CHAPTER 1: Inventing the Cold Chain

1. Henry David Thoreau, *Walden* (1854; Princeton: Princeton University Press, 2004), 294.

2. Ibid., 295.

3. Ibid., 297–98.

4. W. T. Wood & Co., "How to Harvest Ice" (Arlington, MA, 1896), 2–7. See also "Wyeth's Patent on a Manner of Cutting Ice," 18 Mar. 1829, excerpted in Richard O. Cummings, *The American Ice Harvests* (Berkeley: University of California Press, 1949), 145–46. The great similarity between these two descriptions demonstrates how the ice harvesting process remained essentially the same throughout the nineteenth century.

5. Wood & Co., "How to Harvest Ice," 8–11.

6. The best biographical treatments of Frederic Tudor are Gavin Weightman, *The Frozen Water Trade* (New York: Hyperion, 2003); and Carl Seaburg and Stanley Paterson, *The Ice King: Frederic Tudor and His Circle* (Boston: Massachusetts Historical Society, 2003). Most of the biographical details come from Weightman, *Frozen Water Trade*, 16–19. I have examined Tudor's diaries and other papers, but because my time in Boston was limited and Tudor's handwriting is extremely hard to read, I often rely on the work of other historians for Tudor's own words.

7. Philip Chadwick Foster Smith, *Crystal Blocks of Yankee Coldness: The Development of the Massachusetts Ice Trade from Frederick Tudor to Wenham Lake* (Wenham, MA: Wenham Historical Association and Museum, Aug. 1962), 12–13.

8. *Boston Gazette* quoted in Weightman, *Frozen Water Trade*, 37.

9. Seaburg and Paterson, *Ice King*, 31–32; Frederic Tudor, Petition to the Government of Martinique, 1 July 1808, box 29, Tudor Collection II, Frederic Tudor Papers, Historical Collections, Baker Library, Harvard Business School, Boston; Tom Nicholas and Sandra Nicholas, "The Ice King," Harvard Business School Case Study 9-808-094, 17 April 2008, 6.

10. Robert Maclay, "The Ice Industry," in Chauncey Mitchell Depew, *One Hundred Years of American Commerce* (New York: D. O. Haynes, 1895), 2:467.

11. Ibid.

12. Frederic Tudor to John Savage, 2 Oct. 1805, in Nicholas and Nicholas, "Ice King," 15. Also see Weightman, *Frozen Water Trade*, 66–68.

13. Norman E. Borden, Jr., *Dear Sarah: New England Ice to the Orient and Other Incidents from the Journals of Captain Charles Edward Barry to His Wife* (Freeport, ME: Bond Wheelwright, 1966), 42–43.

14. James Parton, *Captains of Industry* (Boston: Houghton Mifflin, 1912), 159.

15. Captain W. J. Lewis Parker, "The East Coast Ice Trade of the United States," in *Ice Carrying Trade at Sea*, ed. D. V. Proctor, Monographs and Reports no. 49 (Greenwich, UK: National Maritime Museum Maritime, 1981), 5, 17.

16. Smith, "Crystal Blocks of Yankee Coldness," 16, 14.

17. Cummings, *American Ice Harvests*, 19–21.

18. Henry Hall, "The Ice Industry of the United States," 10, in Census Reports, *Report on Power and Machinery Employed in Manufactures*, 10th Census (Washington, DC, 1888).

19. Weightman, *Frozen Water Trade*, 106–9.

20. Hall, "Ice Industry of the United States," 12.

21. Charles Dickens, *American Notes and Pictures from Italy* (New York: Scribner's, 1900), 97.

22. Frederic Tudor Diaries, 10 Feb. 1830, microfilm reel 1, Baker Library, Harvard Business School, Boston.

23. John W. Damon, *The Havana Ice-House Controversy* (Boston, 1846), 208.

24. Benjamin Waterhouse quoted in Weightman, *Frozen Water Trade*, 154.

25. Jill Sinclair, *Fresh Pond: The History of a Cambridge Landscape* (Cambridge: MIT Press, 2009), 43.

26. Elizabeth David, *Harvest of the Cold Months* (New York: Viking, 1995), 31; Margaret Visser, *Much Depends on Dinner* (New York: Collier, 1988), 288–89.

27. Hippocrates quoted in Thomas Masters, *The Ice Book* (London: Simpkin, Marshall, 1841), 9–10.

28. See, for example, W. W. Hall, "Iced-Water," *Hall's Journal of Health* 7 (1860): 173.

29. Herbert Spencer, *An Autobiography* (London: Harrison and Sons, 1904), 2:398–99.

30. Marilyn Powell, *Ice Cream: The Delicious History* (New York: Overlook Press, 2006), 23.

31. Frederick Tudor to Charles McKensie, 30 March 1826, box 11, Tudor Collection II, Baker Library.

32. Seaburg and Paterson, *Ice King*, 32.

33. Undated memo from 1806, box 31, Tudor Collection II.

34. Weightman, *Frozen Water Trade*, 70–71.

35. Frederic Tudor to L. Lewis, 14 April 1828, box 12, Tudor Collection II.

36. "Ice: and the Ice Trade," *Hunt's Merchants' Magazine* 33 (Aug. 1855): 170–71.

37. Leander Wetherell, "The Ice Trade," in *Report of the Commissioner of Agriculture* (Washington, DC, 1863), 441.

38. Frederic Tudor to Stephen Cabot, 15 Dec. 1820, box 30, Tudor Collection II.

39. Frederic Tudor to John Barnard, 1 Feb. 1821, box 11, Tudor Collection II.

40. *Ice and Refrigeration* 21 (July 1901): 9.

41. Frederic Tudor to Robert Hooper, 22 Jan. 1849, in *Proceedings of the Massachusetts Historical Society , 1855–1858*, Jan. 1856, 56.

42. Wetherell, "The Ice Trade," 439.

43. Sinclair, *Fresh Pond*, 39–41.

44. Hall, "Ice Industry of the United States," 3.

45. *Chamber's Edinburgh Journal* 11 (6 Jan. 1849): 93.

46. Hall, "Ice Industry of the United States," 5.

47. Weightman, *Frozen Water Trade*, 212.

48. "Ice: and the Ice Trade," 170–71.

49. Weightman, *Frozen Water Trade*, 126.

50. Roger Thévenot, *A History of Refrigeration throughout the World*, trans. J. C. Fidler (Paris: International Institute of Refrigeration, 1979), 67.

51. Smith, "Crystal Blocks of Yankee Coldness," 13, 35.

52. Colesworthey Grant, *Anglo-Indian Domestic Life*, 2nd ed. (Calcutta: Thacker, Spink, 1862), 36–37. Emphasis in original.

53. Ibid., 123.

54. *Calcutta Gazette*, quoted in Weightman, *Frozen Water Trade*, 143–44.

55. W. W. Bunting, "The East India Ice Trade," in Proctor, *Ice Carrying Trade at Sea*, 22.

56. David Dickason, "The Nineteenth-Century Indo-American Ice Trade: An Hyperborean Epic," *Modern Asian Studies* 25 (Feb. 1991): 69–70.

57. Bunting, "The East India Ice Trade," 22.

58. Unnamed London newspaper quoted in *Daily National Intelligencer*, 12 June 1845.

59. *Illustrated London News*, 11 July 1863, 51.

60. Hall, "Ice Industry of the United States," 3.

61. Sarah Mytton Maury, *An Englishwoman in America* (London: Thomas Richardson and Son, 1848), 200.

62. Barry Donaldson and Bernard Nagengast, *Heat and Cold: Mastering the Great Indoors* (Atlanta: American Society of Heating, Refrigerating and Air-Conditioning Engineers, 1994), 46.

63. Hall, "Ice Industry of the United States," 3.

64. Monica Ellis, *Ice and Icehouses through the Ages: With a Gazetteer for Hampshire* (Southampton, UK: Southampton University Industrial Archeology Group, 1982), 31–32.

65. *Ice Trade Journal* 19 (June 1896): 6; Sylvia P. Beamon and Susan Roaf, *The Ice-Houses of Britain* (London: Routledge, 1989), 48.

66. David, *Harvest of the Cold Months*, 241–42.

67. John Burroughs, "Our River," *Scribner's Monthly* 20 (Aug. 1880): 483–84.

68. Maclay, "Ice Industry," 467.

69. Burroughs, "Our River," 484.

70. Hall, "Ice Industry of the United States," 24.

71. Gifford Wood Company, *Natural Ice Machinery*, Catalog No. 60, ca. 1924, 3.

72. Burroughs, "Our River," 484–85. After the turn of the century, these ice elevators came to resemble elevators for people, hoisting blocks of ice that weighed many tons straight up to the top of the icehouses.

73. J. T. Trowbridge, *Lawrence's Adventures among the Ice-Cutters, Glass-Makers, Coal Miners, Iron-Men, and Ship-Builders* (Boston: Fields, Osgood, 1871), 25.

74. W. P. Blake, "Mining and Storing Ice," *Journal of the Franklin Institute* 86 (Nov. 1883): 364.

75. Tom Lewis, *The Hudson: A History* (New Haven: Yale University Press, 2005), 244–45.

76. Marc Ferris, "Ice Harvesting," *The Encyclopedia of New York City*, ed. Kenneth T. Jackson (New Haven: Yale University Press, 1995), 579.

77. Maclay, "Ice Industry," 468.

78. Hall, "Ice Industry of the United States," 27.

79. *Kennebec Journal*, 29 July 1961.

80. L. C. Ballard, "Maine Ice Industry," in *Fifth Annual Report of the Bureau of Industrial and Labor Statistics for the State of Maine* (Augusta: Burleigh & Flynt, 1892), 164–65.

81. Louis Clinton Hatch, *Maine: A History* (New York: American Historical Society, 1919), 3:684.

CHAPTER 2: The Long Wait for Mechanical Refrigeration

1. Information about the Chicago World's Fair fire comes from Chicago newspaper accounts of the tragedy published in the *Chicago Record*, the *Chicago Inter Ocean*, the *Chicago Daily News*, the *Chicago Evening Post*, and the *Chicago Tribune* between 10 and 20 July 1893. See also, Jonathan Rees, "'I Did Not Know . . . Any Danger Was Attached': Safety Consciousness in the Early American Ice and Refrigeration Industries," *Technology and Culture* 46 (July 2005): 541–43.

2. *Chicago Inter Ocean*, 18 July 1893, 2.

3. Ibid., 11 July 1893, 2.

4. Benevolent Association of the Paid Firemen of Chicago, *A Syntopical History of the Chicago Fire Department* (Chicago, 1908), 75. A coroner's jury did refer two officials from the Hercules Iron Works, one World's Fair official responsible for fire safety, and one officer from the Chicago Fire Department to a grand jury to be investigated for criminal negligence. No charges were ever brought. The fact that each man could be judged partly responsible probably saved them all from prosecution. The officials of the Hercules Iron Works should have installed the thimble on their smokestack. The World's Fair official should have forced Hercules to do so, and the fire department (in particular, Captain Fitzpatrick) should have checked to see if the fire had spread to the bottom of the smokestack before sending men up into the tower.

Even though it was not legally responsible for the fire, the Hercules Iron Works lost about $250,000 because it had little insurance on the building. This loss led to a reorganization of the company. By July 1896, the company was in receivership. *Ice and Refrigeration* 5 (Aug. 1893): 91, and ibid. 11 (July 1896): 36d.

5. *Chicago Inter Ocean*, 11 July 1893, 1. My friend Andrew Pearson insists that an ammonia explosion would be impossible. He writes: "The tragedy at the Cold Storage building of the Chicago World's Fair on July 10 1893 was caused by a combination of human error in designing the building, in executing its construction and commissioning and in responding to the initial fire alarm. It is clear from a detailed analysis of the spread of the fire that the use of ammonia in the refrigeration system was not a contributory factor and that the presence of ammonia on site did not hamper the fire-fighting or rescue efforts, nor did it significantly accelerate the destruction of the building once the fire-fighting had been abandoned." "Lessons Learned from the Cold Storage Fire at the World's Fair, Chicago, 1893," paper presented at the American Society of Heating Refrigeration and Air-Conditioning Engineers meeting, Chicago, 2009.

Another possibility is that it was not the gas but the lubricating oil in the pipe that caught on fire. See *Ice and Refrigeration* 18 (Sept. 1898): 157. Regardless of the cause of the actual fire, the important point here is that the public developed concerns over the safety of mechanical refrigeration (real or imagined) as a result of this tragedy.

6. Mark Twain, *Life on the Mississippi* (New York: Harper, 1901), 285.

7. *Ice and Refrigeration* 3 (Dec. 1892): 427.

8. Ibid., 428. Visitors to the fair could still travel on an "Ice Railway," supplied by machines manufactured by the De La Vergne Refrigerating Machine Company of New York City. The *Ice Trade Journal* 17 (Oct. 1893): 5, reported: "The thousands of visitors all pronounce the charm of sleighing on a real bed of snow in midsummer worth the entire cost of a trip to the fair."

9. *Milwaukee Journal*, 4 Nov. 1894.

10. De La Vergne Refrigerating Machine Co., *Mechanical Refrigeration and Ice Making* (New York, 1898), 9.

11. A. J. Wallis-Tayler, *Refrigerating and Ice-Making Machinery*, 2nd ed. (London: Crosby-Lockwood, 1897), 252. My technical explanation of refrigeration comes from reading numerous refrigeration engineering texts like this one and from conversations with two people whose knowledge of this subject is much better than mine. Professor Douglas Reindl of the Industrial Refrigeration Consortium at the University of Wisconsin—Madison invited me to an ammonia refrigeration training class in March 2003. Andrew Pearson of Star Refrigeration Ltd. of Glasgow Scotland helped me with technical issues and also shared his enormous knowledge of refrigeration history with me. Whatever mistakes that may remain are my own.

12. Willis R. Woolrich, "The History of Refrigeration: 220 Years of Mechanical and Chemical Cold, 1748–1968," *ASHRAE Journal* 11 (July 1969): 37.

13. Wallis-Tayler, *Refrigerating and Ice-Making Machinery*, 40.

14. Frick Company, *Seventy-Five Years of Progress* (1928; rpt., Hagerstown, MD: Hagerstown Bookbinding and Print, 1944), 37.

15. Alexander Twining, *The Manufacture of Ice on a Commercial Scale . . .* (New Haven: Thomas J. Stafford, 1857), 3.

16. Roger Thévenot, *A History of Refrigeration throughout the World*, trans. J. C. Fidler (Paris: International Institute of Refrigeration, 1979), 40; Alexander C. Twining, *The Fundamental Ice-Making Invention*, Committee on Patents, U.S. House of Representatives (Washington, DC: H. Polkinhorn, 1870), 1–4.

17. John Gorrie, "Improved Process for the Mechanical Artificial Production of Ice," U.S. Patent 8,080, issued 6 May 1851, 1.

18. Ross E. Hutchins, "Apalachicola: Birthplace of Mechanical Refrigeration," *Popular Mechanics* 110 (Sept. 1958): 122.

19. Barry Donaldson and Bernard Nagengast, *Heat and Cold: Mastering the Great Indoors* (Atlanta: American Society of Heating, Refrigerating and Air-Conditioning Engineers, 1994), 119–20, 124.

20. *Ice and Refrigeration* 3 (Oct. 1892): 270.

21. J. E. Siebel, *Compend of Mechanical Refrigeration and Engineering*, 7th ed. (Chicago: Nickerson and Collins, 1906), 159.

22. U.S. Environmental Protection Agency, "Hazards of Ammonia Releases at Ammonia Refrigeration Facilities (Update)," Chemical Safety Alert, EPA 550-F-01-009, Aug. 2001, 2.

23. See, for example, the 1889 accident described in Gavin Weightman, *The Frozen Water Trade* (New York: Hyperion, 2003), 230.

24. Rees, "'I Did Not Know . . . Any Danger Was Attached,'" 544–60. Much of the subsequent discussion of safety in the mechanical refrigeration industry is drawn from this article.

25. Thomas Shipley, "Recollections of the Ice Machine Industry," *Ice and Refrigeration* 51 (Nov. 1916): 163.

26. Writing in 1900, the engineer M. C. Bersch claimed that these purchasers would pick the larger machine over a smaller, more efficient machine of the same type 90 percent of the time. M. C. Bersch, "To Prove Efficiency of Ice Machines without a Trial," *Ice and Refrigeration* 18 (April 1900): 331.

27. W. D. Humphrey to F. Sparre, "Du Pont Refrigerants: A Comparative Study of Our Activities in This Field . . . ," 30 March 1931, Records of E. I. Du Pont de Nemours & Company, series II, pt. 2, box 1036, Hagley Museum and Library, Wilmington, DE. When Frigidaire was deciding on the proper refrigerant to use for its first household refrigerators, it investigated 600 possible substances. Alexander Stevenson, Jr., *Report on Domestic Refrigerating Machines, 1923–1925*, transcribed by Anne Marie Nagengast, accessed 5 Nov. 2010 at www.ashrae.org/aboutus/page/150, 258; this report is also available online as a PDF file via a Google search.

28. Other less successful refrigerants used in the quest for commercially viable mechanical refrigeration included ether, methylic chloride, air, nitrous oxide, methylmine, and chymogene. De La Vergne Refrigerating Machine Co., *Mechanical Refrigeration* (catalog), 1883, 8.

29. De La Vergne Refrigerating Machine Co., *The De La Vergne Refrigerating Machine Company of New York City* (catalog), 2nd ed. (New York, 1887), 17–19.

30. Woolrich, "History of Refrigeration," 35.

31. Thévenot, *History of Refrigeration throughout the World*, 46.

32. U.S. Department of Commerce, Bureau of the Census, "Manufactured Ice," *Fourteenth Census of the United States Manufactures, 1919* (Washington, DC: GPO, 1922), 12.

33. Gideon Harris and Associates, *Audel's Answers on Refrigeration and Ice Making* (New York: Theo. Audel & Co., 1914), 393–94; Bernard Nagengast to Jonathan Rees (e-mail), 8 Dec. 2010.

34. Pictet Artificial Ice Company, *New Machine for the Production of Artificial Ice and Cold*, ca. 1880.

35. A. B. Pearson, "Carbon Dioxide: New Uses for an Old Refrigerant," International Congress of Refrigeration, Washington, DC, 2003, 2, 5.

36. Harris, *Audel's Answers on Refrigeration and Ice Making*, 371.

37. *Kroeschell Brothers Ice Machine Company*, 5th catalog, National Museum of America History (NMAH), Refrigeration and Miscellaneous Pamphlets, Washington, DC. There were various ways to manufacture ammonia in the late nineteenth century, but the one most often used was as a by-product of coke manufacture or of the manufacture of illuminating gas. In 1905, the German chemist Fritz Haber developed a way to create ammonia synthetically at high temperatures. This proved essential for the German war effort during World War I because an allied boycott had cut off Germany's coal supply and it needed ammonia to make explosives. The widespread adoption of the Haber process throughout the industrialized world reduced the price of ammonia significantly. Synthetic ammonia entered the American market about 1920. See J. Grossman, *Ammonia and Its Compounds* (New York: D. Van Nostrand, 1907), 11–29; Albert Stwertka, *Guide to the Elements*, rev. ed. (Oxford: Oxford University Press, 1998), 43–44; Chas Herter, "Joint Meeting of New York Chapters . . . ," *Ice and Refrigeration* 69 (Nov. 1925), 280.

38. Carbon dioxide compression equipment always came equipped with safety valves in order to relieve the high pressure and prevent explosions. While ammonia gas is toxic, carbon dioxide can kill only by preventing air from reaching the lungs, an event that safety equipment designed to control release of the gas could always prevent.

39. De La Vergne Refrigerating Machine Co., *Mechanical Refrigeration* (catalog), 1890, 21.

40. *Manufacturer and Builder* 12 (Sept. 1880): 201.

41. Oscar Edward Anderson, Jr., *Refrigeration in America* (Princeton: Princeton University Press, 1953), 105.

42. Hans-Liudger Dienel, *Linde: History of a Technology Corporation, 1879–2004* (New York: Palgrave Macmillan, 2004), 32–33, 41, 51, 54.

43. Boyle Ice Machine Co., *Ice Machines and Refrigerating Apparatus* (Chicago, ca. 1880).

44. *Ice and Refrigeration* 7 (Oct. 1894): 243.

45. In 1903, the Chronicle Company of New York published a compendium covering 28 years of fire loss statistics in the United States by type of building. It listed 159 fires in artificial ice plants between the years 1875–1902. At the end of this period, there were only 787 such establishments in the entire country. *The Chronicle of Fire Tables for 1903* (New York: Chronicle, 1902), 273, 461, 464. After the Chronicle Company's last compilation of fire statistics in 1902, nobody seems to have kept exact numbers. John E. Starr, "Accidents in Refrigerating Plants," *A.S.R.E. Journal* 3 (March 1917): 5.

46. Charles D. Havenstrite, "Purchasing a Refrigerating Plant," *Cold Storage and Ice Trade Journal* 34 (Oct. 1907): 34.

47. George L. Reuschline, "What Becomes of the Ammonia in Refrigerating Systems?" *A.S.R.E. Journal* 4 (Sept. 1919): 163.

48. Charles C. Dominge and Walter O. Lincoln, *Fire Insurance Inspection and Underwriting* (New York: Spectator, 1920), 26.

49. *Ice and Refrigeration* 67 (Oct. 1924): 239.

50. See, for example, Kroeschell Brothers Ice Machine Co., *Modern Refrigerating & Ice Making Machinery*, 1901, 60.

51. Anderson, *Refrigeration in America*, 97.

52. Harry Miller, *Halls of Dartford, 1785–1985* (London: Hutchinson Benham, 1985), 71–72.

53. Ibid., 72.

54. Donaldson and Nagengast, *Heat and Cold*, 134.

55. “Refrigeration,” Texas State Historical Association, The Handbook of Texas Online, www.tshaonline.org.

56. Thévenot, *History of Refrigeration throughout the World*, 42.

57. John K. Brown, *The Baldwin Locomotive Works, 1831–1915* (Baltimore: Johns Hopkins University Press, 1995), xxvi–xxvii.

58. Stephen P. Walker, *Lemp: The Haunting History*, 2nd ed. (St. Louis: Mulligan, 1988), 10–11.

59. De La Vergne Refrigerating Machine Co., *Mechanical Refrigeration*, 1883, 43, 61.

60. Anderson, *Refrigeration in America*, 86–91.

61. J. C. Goosmann, “History of Refrigeration,” *Ice and Refrigeration* 69 (Oct. 1925): 203–4.

62. *Ice and Refrigeration* 21 (Sept. 1901): 90.

63. *Frank Leslie's Illustrated Newspaper*, 14 Feb. 1891, 30.

64. Goosmann, “History of Refrigeration,” 267.

65. De La Vergne Refrigerating Machine Co., *De La Vergne Refrigerating and Ice-Making Machinery*, 1908, 3.

66. *Ice Trade Journal* 9 (March 1886): 4.

67. Frick Co., *“Eclipse” Refrigerating Machines* (Waynesboro, PA, 1890), 13.

68. *Cold Storage* 3 (Feb. 1900): 34.

69. Iltyd Redwood, *Theoretical and Practical Ammonia Refrigeration* (New York: Spon & Chamberlain, 1902), 1.

70. Ibid., 1–2.

71. George M. Brill, “Advantages of Artificial Refrigeration,” *Cold Storage* 7 (Jan. 1902): 87.

72. *Cold Storage* 2 (Sept. 1899): 49.

73. U.S. Department of State, *Refrigerators and Food Preservation in Foreign Countries* (Washington, DC: U.S. Government Printing Office, 1891): 62, 81.

74. *Ice and Cold Storage* 1 (July 1898): 77.

75. Miller, *Halls of Dartford*, 66.

76. W. R. Woolrich, “The Trends of Refrigeration in Great Britain,” *Ice and Refrigeration* 117 (Aug. 1949): 40.

77. U.S. Department of Commerce, *Ice-Making and Cold Storage Plants in South America*, *Trade Information Bulletin* 209, 10 March 1924, 2, 3.

78. Department of State, *Refrigerators and Food Preservation in Foreign Countries*, 130.

79. Dienel, *Linde*, 59–60. According to Dienel, Linde did erect other ice works around Germany starting in 1896.

80. Thévenot, *A History of Refrigeration throughout the World*, 131, 133–34.

81. *Ice Trade Journal* 17 (Oct. 1893): 3.

82. *Cold Storage* 4 (Dec. 1900): 141.

83. Charles A. L. Loney, “Refrigeration at Japanese Exposition,” *Ice and Refrigeration* 26 (Jan. 1904): 57–58.

84. *Cold Storage* 6 (Oct. 1901): 91.

85. Donaldson and Nagengast, *Heat and Cold*, 149–50. See also David L. Fiske, “The Time I Speak Of,” *Refrigerating Engineering* 34 (Dec. 1934): 287.

86. Anderson, *Refrigeration in America*, 97–98; Bernard Nagengast, “Electrical Refrigerators Vital Contribution to Households,” *100 Years of Refrigeration*, A Supplement to *ASHRAE Journal* (Nov. 2004): S3-S4. Many thanks to Bernard Nagengast for sending me his important article in PDF format.

87. *Ice Trade Journal* 17 (Sept. 1893): 5.

88. *New York Tribune*, 7 March 1897, A14.

89. *Omaha Daily Bee*, 4 April 1897, 16.

90. C. Linde, "The Refrigerating Machine of Today," *The Engineer* 76 (22 Dec. 1893): 598.

91. *New York Tribune*, 13 May 1900, A5.

92. *Cold Storage and Ice Trade Journal* 17 (Feb. 1907): 34.

93. Anderson, *Refrigeration in America*, 101.

94. John Meyer and R. L. Lloyd, "Electrical Refrigeration in Philadelphia," *Electrical World* 51 (1908): 1210.

95. Harris, *Audel's Answers on Refrigeration and Ice Making*, 14.

96. Anderson, *Refrigeration in America*, 96.

97. *Ice Trade Journal* 17 (June 1894): 4; H. W. Bahrenberg, "The President's Address," 17 Nov. 1910, in *Proceedings of the Second Annual Meeting of the Natural Ice Association of America*, (New York: Natural Ice Association of America, 1910), 10.

98. Hugh Meloy, "Manufactured Ice," in U.S. Bureau of the Census, *Census of Manufactures 1905: Slaughtering and Meat Packing, Manufactured Ice and Salt*, Bulletin 83 (Washington, DC: GPO, 1907), 45.

99. H. D. Pownall, "The Use of Refrigerating Machinery," *Power* 32 (1910): 1048–49. Clipping from Roy Eilers Collection, NMAH.

100. Thévenot, *A History of Refrigeration throughout the World*, 121–23.

101. *Electrical World* 56 (1 Dec. 1910): 1268.

102. A. J. Authenrieth, "Ice—Its History and Importance," in Middle West Utilities, *Ice: A Handbook of Ice, 1927*, 7.

103. *Electrical World* 56 (1 Dec. 1910): 1268.

CHAPTER 3: The Decline of the Natural Ice Industry

1. The first three sections of this chapter are adapted from Jonathan Rees, "What's Left at the Bottom of the Glass: The Quest for Purity and the Development of the American Natural Ice Industry," in *Food Chains: From Farmyard to Shopping Cart*, ed. Warren Belasco and Roger Horowitz (Philadelphia: University of Pennsylvania Press, 2009), 108–25.

2. Julius Adams, "On the Pollution of Rivers," in *Report of the Commission of Engineers on the Water Supply of Philadelphia* (Philadelphia: E. C. Markley, 1875), 65.

3. Ibid., 51.

4. *Trenton State Gazette*, 13 July 1882, 2.

5. "C.M.C.," *Philadelphia Public Ledger Supplement*, 13 Jan. 1883.

6. *Ice Trade Journal* 7 (March 1884): 2.

7. The Stilwell-Bierce & Smith-Vaile Co., *The "Victor" Ice and Refrigerating Machines*, 1899, National Museum of American History, Refrigeration and Miscellaneous Pamphlets, Washington, DC.

8. *New York Times*, 10 Nov. 1901.

9. Hollis Godfrey, "The City's Ice," *Atlantic Monthly* 104 (July 1909): 121.

10. Henry David Thoreau, *Walden* (1854; Princeton: Princeton University Press, 2004), 296.

11. *Ice Trade Journal* 5 (Dec. 1881): 2.

12. Theron L. Hiles, *The Ice Crop: How to Harvest, Store, Ship and Use Ice* (New York: Orange Judd, 1893), 15.

13. W. J. Snyder, "Ice from a Scientific Standpoint," *Cold Storage and Ice Trade Journal* 39 (April 1910): 37.

14. *Ice and Refrigeration* 47 (July 1914): 12.

15. Joseph C. Jones, *American Ice Boxes* (Humble, TX: Jobeco Books, 1981), 29.

16. *Detroit Tribune*, quoted in *Cold Storage* 10 (Aug. 1903): 64.

17. *Cold Storage and Ice Trade Journal* 31 (June 1906): 60.

18. Paul Angle, "Ice Wagons," *Chicago History* 5 (Spring 1959): 250.

19. De La Vergne Refrigerating Machine Co., *Ice-Manufacture*, 1892, 13–14.

20. Mark Twain, *Life on the Mississippi* (New York: Harper, 1901), 286.

21. *Ice and Refrigeration* 1 (Dec. 1891): 338.

22. *Cold Storage and Ice Trade Journal* 39 (June 1910): 42.

23. *Cold Storage and Ice Trade Journal* 34 (Sept. 1907): 38.

24. *Detroit Tribune*, quoted in *Cold Storage* 10 (Aug. 1903): 64.

25. *Cold Storage* 8 (Nov. 1902): 212.

26. R. H. Hutchings and A. W. Wheeler, "An Epidemic of Typhoid Fever Due to Impure Ice," *American Journal of the Medical Sciences* 126 (Oct. 1903): 680–84. Modern research has confirmed these findings. For example, a 1985 study (conducted to test the safety of ordering cocktails in underdeveloped countries) found that 20 percent of typhoid bacteria survived in ice cubes after a week. See Lynn Dickens, Herbert L. DuPont, and Philip C. Johnson, "Survival of Bacteria Enteropathogens in the Ice of Popular Drinks," *Journal of the American Medical Association* 253 (7 June 1985): 3141–43.

27. Jonathan M. Zenilman, "Typhoid Fever," *Journal of the American Medical Association* 278 (10 Sept. 1997): 848.

28. Arthur Lederer, "The Relation of Public Water Supplies to General and Specific Mortalities in Cities," *American Journal of Public Hygiene* 20 (May 1910): 303.

29. Nelson Manford Blake, *Water for the Cities* (Syracuse, NY: Syracuse University Press, 1956), 260–61.

30. Michael McCarthy, *Typhoid and the Politics of Public Health in Nineteenth-Century Philadelphia* (Philadelphia: American Philosophical Society, 1987), 11.

31. *Frank Leslie's Illustrated Newspaper*, 16 Sept. 1882, 51.

32. William R. D. Blackwood, "Ice—How to Obtain It Pure," *Ice and Refrigeration* 5 (Dec. 1893): 390.

33. *New York Times*, 31 May 1903.

34. For example, Glenwood Ice & Coal Co., "Subject: Purity of Natural Ice," in *Proceedings of the Natural Ice Association of America*, New York, 12–13 April 1910, 98.

35. John C. Sparks, "Bacteria in Ice," ibid., 40.

36. *Chicago Chronicle*, quoted in *Cold Storage* 10 (July 1903): 16.

37. For example, *Ice and Refrigeration* 47 (July 1914): 12.

38. Upton Sinclair, *The Jungle* (New York: Doubleday, 1906), 33.

39. Trade journals such as *Ice and Refrigeration* and the *Ice Trade Journal* reported on conditions of the ice industry in various markets but did not report price data. The assessment that follows is based on extensive reading of the available trade literature; exact figures are not available.

40. *Philadelphia North American*, 29 May 1893.

41. Much of this paragraph and the remainder of this section comes from Jonathan Rees, "The Natural Price of Natural Ice," *Business and Economic History On-Line* 6 (2008), www.thebhc.org/publications/BEHonline/2008/rees.pdf.

42. This would explain why the government did not add ice to its cost of living index until 1934. See Thomas Stapleford, *The Cost of Living in America: A Political History of Economic Statistics, 1880–2000* (New York: Cambridge University Press, 2009), 162.

43. *Cold Storage and Ice Trade Journal* 32 (Aug. 1906): 38.

44. *Cold Storage* 1 (April 1899): 3.

45. *New York Tribune*, 25 Jan. 1894, 3.

46. *Brooklyn Daily Eagle*, 17 July 1876.

47. *New Orleans Daily Picayune*, 15 July 1890.

48. Americus [pseud.], "Ice and Ice Making," *Ice and Refrigeration* 14 (Jan. 1898): 99.

49. *Ice and Refrigeration* 4 (May 1893): 373.

50. *Ice Trade Journal* 21 (May 1898): 1.

51. Oscar Edward Anderson, Jr., *Refrigeration in America* (Princeton: Princeton University Press, 1953), 116.

52. Ibid., 40.

53. Edward T. O'Donnell, "The Dawn of New York's Ice Age," *New York Times*, 31 July 2005.

54. Henry Hall, "The Ice Industry of the United States," 4, 27 in *Census Reports: Report on Power and Machinery Employed in Manufactures*, Tenth Census (Washington, DC: GPO, 1888).

55. *New York Tribune*, 13 Dec. 1891, 16.

56. Owen Wilson, "The Admiral of the Atlantic Coast," *World's Work* 13 (April 1907): 8719.

57. Gavin Weightman, *The Frozen-Water Trade* (New York: Hyperion, 2003), 237–39.

58. Anderson, *Refrigeration in America*, 117.

59. O'Donnell, "Dawn of New York's Ice Age."

60. *Ice and Refrigeration* 6 (Jan. 1894): 13.

61. Jill Jonnes, *Conquering Gotham: Building Penn Station and Its Tunnels* (New York: Penguin Books, 2007), 78.

62. *Cold Storage* 4 (Aug. 1900): 38; ibid. 4 (Oct. 1900): 84.

63. A. R. Norton, "Few Persons Know How to Make Wild or Tame Ice Do Its Duty," *New York Tribune*, 25 April 1915, 5.

64. Natural Ice Association of America, *The Handwriting on the Wall*, New York, ca. 1913, box B-13, Lake Mohonk Mountain House Collection, Hagley Museum and Library, Wilmington, DE.

65. *New York Tribune*, 13 May 1894, 25.

66. Triumph Ice Machine Co., *Manufacturers of Ice and Refrigerating Machinery*, 1901–2 catalog, Cincinnati, OH, 45.

67. Horace B. Drury, "Production and Capacity Control in the Ice Industry under the NRA," National Recovery Administration, Trade Practices Studies Section, March 1936, 56.

68. Adam Ley, "Manufactured Ice a Novelty," *Ice* 4 (Feb. 1909): 205.

69. *Manufacturer and Builder* 11 (July 1879): 147.

70. *Ice Trade Journal* 20 (Aug. 1896): 6.

71. Ibid., 10.

72. Ibid. (Dec. 1896): 1.

73. *Cold Storage* 1 (April 1899): 3.

74. H. W. Bahrenburg, "What the Dealer in Natural Ice Must Do to Retain His Position in the Industry," *Ice and Refrigeration* 37 (Dec. 1909): 241.

75. *Cold Storage and Ice Trade Journal* 40 (Dec. 1910): 67.
76. *Ice and Refrigeration* 37 (Oct. 1909): 123.
77. *Ice and Refrigeration* 47 (July 1914): 12.
78. Anderson, *Refrigeration in America*, 107; *Ice and Refrigeration* 54 (March 1918): 172.
79. Jennie Everson, *Tidewater Ice of the Kennebec River* (Freeport, ME: Bond Wheelwright, 1970), 141.
80. Charles H. Ehrenfeld and Ralph E. Gibbs, *Water for Ice-Making and Refrigeration* (Chicago: Nickerson & Collins, 1928), 24.

CHAPTER 4: Refrigerated Transport Near and Far

1. James Harvey Young, *Pure Food: Securing the Federal Food and Drugs Act of 1906* (Princeton: Princeton University Press, 1989), 135–37.
2. *Washington Post*, 24 Jan. 1899.
3. Artemus Ward, "Cold Storage," in *The Grocer's Encyclopedia* (New York, 1911), 181.
4. Theodore Dreiser, "Fruit-Growing in America," *Harper's Monthly Magazine* 101 (Nov. 1900): 860.
5. W. J. Stelpflug, "The Food Industry and the Part That Refrigeration Plays in It," *Analysts Journal* 6 (Fourth Quarter, 1950): 37.
6. According to an online Eugene O'Neill archive, the play's screenwriter, Dudley Nichols, "recorded that O'Neill had wished to recall the Biblical quotation by his use of the archaic verb form, but that he also wanted to suggest the bawdy story of the husband who called upstairs to his wife, 'Has the iceman come yet?' The answer: 'No, but he's breathing hard.'" See "The Door and the Mirror: The Iceman Cometh," eONeill.com, www.eoneill.com/library/contour/mirror/iceman.htm, accessed 3 May 2010.
7. "The Jobs of Yesteryear: Obsolete Occupations," National Public Radio, www.npr.org/templates/story/story.php?storyId=124251060, accessed 3 May 2010.
8. David Margolick, *Beyond Glory: Joe Louis vs. Max Schmeling, and a World on the Brink* (New York: Random House, 2005), 60; Bill Hanks, "Red Grange: The Greatest College Football Player Ever," Associated Content, www.associatedcontent.com/article/514299/red_grange_the_greatest_college_football.html, accessed 25 Oct. 2010.
9. Paul Angle, "Ice Wagons," *Chicago History* 5 (Spring 1959): 250.
10. Walter R. Sanders, *Selling Ice: A Complete Treatise on the Subject . . .* (Chicago: Nickerson and Collins, 1922), 29.
11. Anne Francis Miksinski, "The Quiet Revolution: The History and Effects of Domestic Refrigeration in America; From Ice to Mechanical Refrigerants," M.A. thesis, George Washington University, 1970, 14.
12. *Ice and Refrigeration* 24 (April 1903): 156.
13. For example, the City Delivery Company, noted in Sanders, *Selling Ice*, 56.
14. Sanders, *Selling Ice*, 42.
15. Utica Ice Company, quoted in Dewey D. Hill and Elliot R. Hughes, *Ice Harvesting in Early America* (New Hartford, NY: New Hartford Historical Society, 1977), 28.
16. J. F. Bell & Sons, "Ice Delivery System," ca. 1915, Household Management Collection, Schlesinger Library, Harvard University, Cambridge.
17. J. H. Walsh, *A Manual of Domestic Economy Suited to Families Spending from £150 to £1500 a Year* (London: Routledge, 1874), 441.
18. *The Lancet*, quoted in *Ice and Refrigeration* 27 (July 1904): 26.

19. *Ice and Refrigeration* 27 (Aug. 1904): 72.

20. Ibid., 18 (Jan. 1900): 17.

21. Richard J. Orsi, *Sunset Limited: The Southern Pacific Railroad and the Development of the American West, 1850–1930* (Berkeley: University of California Press, 2005), 331.

22. Frederic Tudor, unaddressed letter, copy #28, Nov. 1824, box 11, Tudor Collection II, Frederic Tudor Papers, Historical Collections, Baker Library, Harvard Business School, Boston.

23. David McCullough, *The Greater Journey: Americans in Paris* (New York: Simon and Schuster, 2011), 210.

24. Thomas Mort, "On the Preservation of Food by Freezing, and the Bearing It Will Have on the Pastoral and Agricultural Interests of Australia," 17 July 1875, in *Journal of the Agricultural Society of New South Wales* 2 (Nov. 1875): 255.

25. Richard Perren, *The Meat Trade in Britain, 1840–1914* (London: Routledge and Kegan Paul, 1978), 81.

26. Mort, "On the Preservation of Food," 247–49.

27. Ibid., 250.

28. "Report of Directors submitted to the General Meetings of the Shareholders of the New South Wales Fresh Food & Ice Co.," 30 Aug. 1878, Reel CY 2901, Mort Papers, State Library of New South Wales, Sydney, Australia.

29. *Sydney Mail*, 28 Sept. 1910.

30. James Jervis, "Thomas Sutcliffe Mort: A National Benefactor," *Royal Australian Historical Society: Journal and Proceedings* 24 (1938): 382.

31. *Monthly Bulletin of the International Association of Refrigeration* 1 (Aug. 1910): 102.

32. *Illustrated London News*, 3 March 1877, 203.

33. Susanne Freidberg, *Fresh: A Perishable History* (Cambridge: Harvard University Press, 2009), 58–61.

34. James T. Critchell and Joseph Raymond, *A History of the Frozen Meat Trade* (London, 1912): 27–29.

35. Roger Thévenot, *A History of Refrigeration throughout the World*, trans. J. C. Fidler (Paris: International Institute of Refrigeration, 1979), 79.

36. *The Standard*, 20 Feb. 1877, 6.

37. *York Herald*, 18 Jan. 1877.

38. David M. Higgins and Dev Gangjee, "'Trick or Treat'? The Misrepresentation of American Beef Exports in Britain during the Late Nineteenth Century," *Enterprise and Society* 11 (June 2010): 209.

39. *The Times* of London, 5 Jan. 1891, 13; R. Duncan, "The Demand for Frozen Beef in the United Kingdom, 1880–1940," *Journal of Agricultural Economics* 12 (June 1956): 82.

40. *New York Tribune*, 21 April 1877, 5.

41. Higgins and Gangjeee, "Trick or Treat?" 210.

42. Richard Perren, *The Meat Trade in Britain, 1840–1914* (London: Routledge and Kegan Paul, 1978), 125.

43. *Ice and Refrigeration* 21 (Sept. 1902): 99.

44. Pedro Berges, "Methods of Transporting by Water Meat under Refrigeration: Development from 1868 to 1913," Third International Congress of Refrigeration, *Proceedings* (Chicago, 1913), 3:328.

45. Critchell and Raymond, *History of the Frozen Meat Trade*, 248, 252.

46. John Soluri, *Banana Cultures: Agriculture, Consumption, and Environmental Change in Honduras and the United States* (Austin: University of Texas Press, 2005), 62.

47. Dan Koeppel, *Banana: The Fate of the Fruit That Changed the World* (New York: Penguin, 2008), 54–56; Sarah Murray, *Moveable Feasts: From Ancient Rome to the 21st Century; The Incredible Journeys of the Food We Eat* (New York: Picador, 2007), 115.

48. Murray, *Moveable Feasts*, 114.

49. Ibid., 120.

50. Soluri, *Banana Cultures*, 37.

51. Baker Ice Machine Company, *Modern Cold Storage*, Bulletin no. 60, ca. 1920, 2.

52. Gideon Harris and Associates, *Audel's Answers on Refrigeration and Ice Making* (New York: Theo. Audel, 1914), 511.

53. Marc Levinson, *The Box* (Princeton: Princeton University Press, 2006), 3.

54. Leo Block, "Mobile Refrigeration: Current Practices and Equipment," *ASHRAE Journal* 8 (June 1966): 69.

55. Lester L. Westling, "A Challenge to Developments in Refrigerated Transports," *ASHRAE Journal* 15 (Dec. 1973): 56.

56. Nicola Twilley, "The Coldscape," *Cabinet* 47 (Fall 2012), http://cabinetmagazine.org/issues/47/twilley.php, accessed 30 Nov. 2012.

57. Murray, *Moveable Feasts*, 35, 38.

58. Robert Heap, "Refrigerated Transport: Progress Achieved and Challenges to Be Met," International Institute of Refrigeration, August 2003.

59. Richard White, *Railroaded: The Transcontinentals and the Making of Modern America* (New York: Norton, 2011), 153–54.

60. Mary Yeager, *Competition and Regulation: The Development of Oligopoly in the Meat Packing Industry* (Greenwich, CT: JAI Press, 1981), 56–57.

61. U.S. Department of State, *Refrigerators and Food Preservation in Foreign Countries* (Washington, DC: GPO, 1891), 74–75, 78.

62. Pierre Desrochers and Hiroko Shimzu, *The Locavore's Dilemma: In Praise of the 10,000 Mile Diet* (New York: Public Affairs, 2012), 73.

63. White, *Railroaded*, 481.

64. Kristin Hoganson, "Meat in the Middle: Converging Borderlands in the U.S. Midwest, 1865–1900," *Journal of American History* 98 (March 2012): 1038.

65. Jackson Lears, *Rebirth of a Nation: The Making of Modern America, 1877–1920* (New York: Harper Collins, 2009), 143.

66. John H. White, Jr., *The American Railroad Freight Car* (Baltimore: Johns Hopkins University Press, 1993), 271. White details these many experiments on pp. 271–73.

67. Ronald M. Labbé and Jonathan Lurie, *The Slaughterhouse Cases: Regulation, Reconstruction, and the Fourteenth Amendment* (Lawrence: University Press of Kansas, 2003), 79.

68. John H. White, *The Great Yellow Fleet: A History of Railroad Refrigerator Cars* (San Marino, CA: Golden West Books, 1986), 23.

69. *Electrical World* 58 (1911): 340. Clipping from the Roy Eilers Collection, National Museum of American History (NMAH), Washington, DC.

70. White, *American Railroad Freight Car*, 283. White describes many other refrigerated railcar designs on pp. 273–84.

71. William Cronon, *Nature's Metropolis: Chicago and the Great West* (New York: Norton, 1991), 234–35.

72. Louis F. Swift with Arthur Van Vlissingen, Jr., *The Yankee of the Yards: The Biography of Gustavus Franklin Swift* (Chicago: A. W. Shaw, 1927), 200.

73. Ibid., 131.

74. Oscar Edward Anderson, Jr., *Refrigeration in America* (Princeton: Princeton University Press, 1953), 59.

75. Thévenot, *History of Refrigeration throughout the World*, 85.

76. Linda Danes-Wingett, "The Ice Car Cometh: A History of the Railroad Refrigerator Car," *San Joaquin Historian* 10 (Winter 1996): 2.

77. *Los Angeles Herald*, 29 March 1910, 3.

78. Shane Hamilton, *Trucking Country: The Road to America's Wal-Mart Economy* (Princeton: Princeton University Press, 2008), 35–36.

79. Perry R. Duis, *Challenging Chicago: Coping with Everyday Life, 1837–1920* (Urbana: University of Illinois Press, 1998), 146–47.

80. *Ice and Refrigeration* 21 (Sept. 1901): 92.

81. Andrew Beahrs, *Twain's Feast: Searching for America's Lost Foods in the Footsteps of Samuel Clemens* (New York: Penguin Press, 2010), 104.

82. Friedberg, *Fresh*, 237.

83. Baker Ice Machine Company, *Refrigeration for the Fisheries Industry*, 1923.

84. Mansel G. Blackford, *Making Seafood Sustainable: American Experiences in Global Perspective* (Philadelphia: University of Pennsylvania Press, 2012), 85.

85. Michael Shawyer and Avillo F. Meina-Pizalli, *The Use of Ice in Small Fishing Vessels* (Rome: U.N. Food and Agriculture Organization, 2003), 1–3.

86. Sebastian Junger, *The Perfect Storm* (New York: Harper Collins, 1997), 40–41.

87. Jane Lear, "Jane Says: Surprise! Frozen Seafood Is Often the Best Choice," *TakePart*, 25 July 2012, http://takepart.com/article/2012/07/25/jane-says-surprise-frozen-seafood-often-best-choice, accessed 26 July 2012.

88. Matthew Morse Booker, "Oyster Growers and Oyster Pirates in San Francisco Bay," *Pacific Historical Review* 75 (Feb. 2006): 73.

89. W. D. Tyler, "Eastern Oysters Are Successful," *Prosperous Washington* (Seattle: Seattle Post-Intelligencer, 1906), 23; Matthew Morse Booker, "Immigrant Biotechnology: Japanese Oysters in the United States," Paper presented at the annual meeting of the Society for the History of Technology, Tacoma, WA, 1 Oct. 2010.

90. A. E. Miller, *Cold Storage Practice* (London: Charles Griffin, 1948), 95.

91. Eugene F. McPike, "The Refrigerator Car: Retrospective and Prospective," *Transactions of the American Society of Refrigerating Engineers* 9 (1915): 128.

92. Rich Stetefeld, *Refrigerated Railway Transport*, Commission on Railway and Steamship Refrigeration of the American Association of Refrigeration, Bulletin no. 1, 25, in Harvey Wiley Papers, box 218, Manuscripts Division, Library of Congress, Washington, DC.

93. *Bulletin of the International Association of Refrigeration* 4 (Jan. 1913): 19–20.

94. John Steinbeck, *East of Eden* (1952; New York: Penguin, 1992), 435–38. Thanks to Gabrielle Petrick for reminding me that Steinbeck's novel turns on refrigerated lettuce.

95. U.S. Department of Agriculture, "Big Loss of Fruits and Vegetables in Refrigerator Cars May Be Decreased" (press release), *Millard County (UT) Chronicle*, 9 Oct. 1919.

96. Steinbeck, *East of Eden*, 437.

97. Harvey Levenstein, *Revolution at the Table: The Transformation of the American Diet* (Berkeley: University of California Press, 2003), 31.

98. Richard H. Hendrickson and Edward S. Kaminski, *Billboard Refrigerator Cars* (Wilton, CA: Signature Press, 2008), 11.

99. *Los Angeles Herald*, 29 Mar. 1910, 3.

100. McPike, "The Refrigerator Car," 130.

101. White, *Great Yellow Fleet*, 18.

102. *Sacramento Bee* excerpted in *Daily Honolulu Press*, 8 March 1886, 4.

103. White, *Great Yellow Fleet*, 18. The need for heaters continued even after refrigerator cars replaced fruit cars. See *Car Builders' Cyclopedia* (New York: Simmons-Boardman, 1928), 166.

104. *Omaha Bee*, 13 Aug. 1874. 4.

105. C. Keith Jordan et al., *Refrigerator Cars: Ice Bunker Cars, 1884–1979* (Norman, OK: Santa Fe Modelers Organization, 1994), 31.

106. Thévenot, *History of Refrigeration throughout the World*, 91.

107. Jordan et al., *Refrigerator Cars*, 39, 44.

108. Harris, *Audel's Answers*, 512.

109. Thévenot, *History of Refrigeration throughout the World*, 91.

110. Alfred D. Chandler, Jr., *The Visible Hand: The Managerial Revolution in America* (Cambridge: Belknap, 1977), 399.

111. Cameron Landon, "The Operations of the Packing Combine," *Perrysburg (OH) Journal*, 31 March 1905, 6.

112. Orsi, *Sunset Limited*, 328–30.

113. White, *American Railroad Freight Car*, 271.

114. Orsi, *Sunset Limited*, 333–35.

115. Ibid., 335, 330–31.

116. *Ice and Refrigeration* 66 (May 1924): 449–50.

117. Anthony W. Thompson, Robert J. Church, and Bruce H. Jones, *PFE: Pacific Fruit Express* (Wilton, CA: Central Valley Road Publications, 1992), 8, 338–39.

118. Thévenot, *History of Refrigeration throughout the World*, 339.

119. *Ice* 7 (Sept. 1910): 57.

CHAPTER 5: The Pleasures and Perils of Cold Storage

1. Harvey W. Wiley, "Food Adulteration and Its Effects," 10–11, typescript, box 198, Harvey W. Wiley Papers, Manuscript Division, Library of Congress, Washington, DC.

2. Harvey W. Wiley, "Use of Cold Storage," U.S. Senate Document no. 486, 61st Cong., 2nd sess., 13 April 1910, 4.

3. Oscar E. Anderson, Jr., *The Health of a Nation: Harvey W. Wiley and the Fight for Pure Food* (Chicago: University of Chicago Press, 1958), 238–39. Wiley did support cold storage labeling laws.

4. *New York Tribune*, 30 Jan. 1910, 3.

5. Horace Greeley, *New York Tribune*, 24 Nov. 1864, quoted in Daniel Somes, *Mr Somes' Inventions for the Preservation of Food . . .* (New York: E. S. Dodge, 1865), 15.

6. Gideon Harris and Associates, *Audel's Answers on Refrigeration and Ice Making* (New York: Theo. Audel & Co., 1914), 550.

7. William Cronon, *Nature's Metropolis: Chicago and the Great West* (New York: Norton, 1991), 231–32.

8. Oscar Edward Anderson, Jr., *Refrigeration in America* (Princeton: Princeton University Press, 1953), 92; W. S. Stair, *The York Organization* (York, PA: York Ice Machine Co., 1942), 20–21.

9. Theron Hiles, *The Ice Crop: How to Harvest, Store, Ship and Use Ice* (New York: Orange Judd, 1893), 67.

10. *Illustrated London News*, 3 March 1877, 203.

11. Edward A. Duddy, *The Cold-Storage Industry in the United States* (Chicago: University of Chicago Press, 1929), 6.

12. Organisation for European Economic Co-operation, *The Cold Chain in the U.S.A.*, pt. 2 (Paris: OEEC, 1952), 143.

13. Hans Peter Henschien, *Packing House and Cold Storage Construction* (Chicago: Nickerson & Collins, 1915), 116, 120–21.

14. David I. Davis, "Cold Storage Construction with Special Reference to the Refrigeration of Meats," Third International Congress of Refrigeration, *Proceedings* (Chicago, 1913), 2:647–48.

15. M. E. Pennington, "Relation of Cold Storage to the Food Supply and the Consumer," *Annals of the American Academy of Political and Social Science* 48 (July 1913): 155–56.

16. *Cyclopedia of Fire Prevention* (Chicago: American Technical Society, 1912), 3:450.

17. Ben Fine and Ellen Leopold, *The World of Consumption* (London: Routledge, 1993), 149.

18. The *Williamsport (PA) Grit*, quoted in Frank A. Horne, "Legislation Affecting Cold Storage and Cold Storage Products," *Ice and Refrigeration* 41 (Nov. 1911): 181.

19. James Harvey Young, *Pure Food: Securing the Federal Food and Drugs Act of 1906* (Princeton: Princeton University Press, 1989), 151.

20. Edgar R. Curry, "The Refrigerated Warehousing Industry—From Early Days to 1949," *Ice and Refrigeration* 116 (Jan. 1949): 21–22.

21. Samuel H. Brubaker, "Timely Thoughts for Cold Storage Interests," *Cold Storage and Ice Trade Journal* 32 (Aug. 1906): 23.

22. Anderson, *Refrigeration in America*, 131.

23. Harris, *Audel's Answers on Refrigeration and Ice Making*, 504–5, 535.

24. Madison Cooper, *Practical Cold Storage* (Chicago: Nickerson & Collins, 1905), 6.

25. Madison Cooper, "Keeping Frost Off Cooling Pipes," *Ice and Refrigeration* 19 (Dec. 1900): 194; also M. R. Carpenter, "Temperature, Humidity and Ventilation in Cold Storage Rooms," *Refrigeration* 37 (Sept. 1925): 34.

26. Harris, *Audel's Answers on Refrigeration and Ice Making*, 539.

27. Milton W. Arrowood, *Refrigeration* (Chicago: American Technical Society, 1917), 228.

28. Cooper, *Practical Cold Storage*, 545.

29. De La Vergne Refrigerating Machine Company, *Refrigerating Plants*, Bulletin no. 144, Aug. 1915, 11.

30. *Cold Storage* 8 (Sept. 1902): 116.

31. *Cold* 4 (June 1913): 157.

32. Americus [pseud.], "Cold Storage of Fruit," *Ice and Refrigeration* 15 (Dec. 1898): 382.

33. Curry, "The Refrigerated Warehousing Industry," 26.

34. *New York Tribune*, 30 Jan. 1910, 3.

35. H. W. Wiley, M.D., "Protection of Fish in Market," 4, Records of the Bureau of Agricultural and Industrial Chemistry, Record Group 97, entry 27, "Articles and Lectures by Harvey W. Wiley," box 3, National Archives II, College Park, MD.

36. Mary Hinman Abel, *Successful Family Life on the Moderate Income* (Philadelphia: Lippincott, 1921), 127–28.

37. Department of Public Markets, City of New York, *Annual Report of the Commissioner of Public Markets* (New York, 1923), 49.

38. U.S. Senate, Committee on Manufactures, *Foods Held in Cold Storage* (Washington: GPO, 1911), 6.

39. Brubaker, "Timely Thoughts for Cold Storage Interests," 23.

40. *Cold Storage* 8 (Dec. 1902): 267.

41. Pennington, "Relation of Cold Storage to the Food Supply," 158.

42. For example, Harvey Wiley in the *Hartford (KY) Herald*, 6 Feb. 1907, 3. He believed at this juncture, before the completion of his experiments, that nothing should be kept in cold storage for over three months.

43. George K. Holmes, *Cold Storage Business Features*, U.S. Department of Agriculture Bureau of Statistics Bulletin 93, 1913, 12–13.

44. U.S. Senate, Committee on Agriculture and Forestry, *Hearing on H.R. 9521*, 66th Cong., 2nd sess. (Washington: GPO, 1920), 59.

45. William Leavitt Stoddard, "How the 'Food Control' Controls Food," *Pearson's Magazine* 31 (April 1914): 467.

46. Joyce Appleby, *The Relentless Revolution: A History of Capitalism* (New York: Norton, 2010), 57.

47. *Cold Storage* 4 (Nov. 1900): 111.

48. *American Review of Reviews* 43 (April 1911), 474.

49. *New York Tribune*, 30 Jan. 1910, 3.

50. Senate Committee on Manufactures, *Foods Held in Cold Storage*, 141.

51. Henschien, *Packing House and Cold Storage Construction*, 118.

52. Lee A. Craig, Barry K. Goodwin, and Thomas Grennes, "The Effect of Mechanical Refrigeration on Nutrition in the United States," *Social Science History* 28 (Summer 2004): 327.

53. Walter E. Clark, *The Cost of Living* (Chicago: A. C. McClurg, 1915), 35.

54. *Ice and Refrigeration* 45 (Oct. 1913): 185.

55. Frank A. Horne, "Cold Storage—Its Capabilities and How to Best Utilize and Extend Them," *American Journal of Public Health* 8 (March 1918): 224.

56. Harris, *Audel's Answers on Refrigeration and Ice Making*, 549–50.

57. *Cold Storage* 5 (June 1901): 159.

58. *Ice and Refrigeration* 63 (Aug. 1922): 93–97.

59. *El Paso Herald*, 30 Oct. 1917, 2.

60. Horne, "Cold Storage," 225–26.

61. I. C. Franklin, "The Effect of the War Upon the Cold Storage Industry," *A.S.R.E. Journal* 6 (Jan. 1920): 244, 248–49.

62. *New York Tribune*, 5 Jan. 1918, 6.

63. Anderson, *Refrigeration in America*, 235.

64. *Saturday Evening Post*, 16 June 1921, in Harvey Wiley Papers, box 208.

65. Senate Committee on Manufactures, *Foods Held in Cold Storage*, 179.

66. Anderson, *Refrigeration in America*, 138–39.

67. Ibid., 137.

68. Frank A. Horne, "The Cold Storage Industry in Relation to the High Cost of Living," 21st New York State Conference of Charities and Correction, 9–11 Nov. 1920, *Proceedings* (Buffalo, 1920), 171.

69. Anderson, *Refrigeration in America*, 236.

70. See, for example, United States Cold Storage, "Computerized Refrigeration Control System," www.uscoldstorage.com/html/comp_ref_sys.html, accessed 2 Jan. 2011.

71. *London Daily News*, 15 Sept. 1886.

72. *The Standard*, 24 Sept. 1891.

73. *Cold Storage* 2 (July 1899): 11.

74. Cooper, *Practical Cold Storage*, 561.

75. *Cold Storage* 3 (May 1900): 102; Sampson Morgan, "Cold Storage and the Christmas Fruit Trade," *Ice and Cold Storage* 9 (Dec. 1906): 247.
76. OEEC, *Cold Chain in the U.S.A.*, pt. 1, 26.
77. *Ice and Cold Storage* 8 (April 1905): 90.
78. *Chambers's Journal* 11 (21 April 1894): 247.
79. *Ice and Cold Storage* 1 (April 1898): 7.
80. *Ice and Refrigeration* 21 (Sept. 1901): 94.
81. *Ice and Cold Storage* 5 (Aug. 1902): 210.
82. *Ice and Cold Storage* 5 (Feb. 1902): 29.
83. Michael Pearson and Jane Lennon, *Pastoral Australia: Fortunes, Failures and Hard Yakka* (Collingswood, Victoria, Australia: Csiro Publishing, 2010), 69.
84. U.S. Department of State, *Refrigerators and Food Preservation in Foreign Countries* (Washington, DC: GPO, 1891), 76, 77.
85. *Monthly Bulletin of the International Association of Refrigeration* 1 (Aug. 1910): 102–4.
86. U.S. Department of Agriculture, *Weather, Crops and Markets* 1 (21 Jan. 1922): 61.
87. Ibid., 3.
88. *Cold Storage* 1 (June 1899): 1.
89. Ibid. 10 (Nov. 1903): 210.
90. *Ice and Refrigeration* 41 (Nov. 1911): 201.
91. Susanne Freidberg, *Fresh: A Perishable History* (Cambridge: Harvard University Press, 2009), 27, 29.
92. Edward E. Slossen, "Science Remaking Everyday Life," *Commercial America* 19 (April 1923): 35.
93. OEEC, *Cold Chain in the U.S.A.*, pt. 2, 38.
94. For example, Hans-Liudger Dienel, *Linde: History of a Technology Corporation, 1879–2004* (New York: Palgrave Macmillan, 2004), 146.
95. A. E. Miller, *Cold Storage Practice* (London: Charles Griffin, 1948), 1.
96. Shane Hamilton, *Trucking Country: The Road to America's Wal-Mart Economy* (Princeton: Princeton University Press, 2008), 121; OEEC, *Cold Chain in the U.S.A.*, pt. 2, 144, 151, 36.
97. Nicola Twilley, "The Coldscape," *Cabinet* 47 (Fall 2012), http://cabinetmagazine.org, accessed 30 Nov. 2012.
98. On peddlers, see Tracey Deutsch, *Building a Housewife's Paradise: Gender, Politics, and American Grocery Stores in the Twentieth Century* (Chapel Hill: University of North Carolina Press, 2010), 28–29.
99. Marc Levinson, *The Great A&P and the Struggle for Small Business in America* (New York: Hill and Wang, 2011), 8, 40, 79, 125.
100. Deutsch, *Building a Housewife's Paradise*, 185–87.
101. Tracie McMillan, *The American Way of Eating: Undercover at Walmart, Applebee's, Farm Fields and the Dinner Table* (New York: Scribner, 2012), 150.
102. Andrew D. Althouse and Carl H. Turnquist, *Modern Electric and Gas Refrigeration* (Chicago: Goodheart-Wilcox, 1937), 366. For more about converting iceboxes to electric household refrigerators, see chapter 7.
103. H. B. Hull, *Mechanical Refrigeration*, 4th ed. (Chicago: Nickerson & Collins, 1933), 50.
104. *ASHRAE Journal* 2 (Sept. 1960): 2.
105. Edward Brand as quoted in Deutsch, *Building a Housewife's Paradise*, 190.

106. Mark Kurlansky, *Birdseye: The Adventures of a Curious Man* (New York: Doubleday, 2012), 163, 185–86.

107. Anderson, *Refrigeration in America*, 283.

108. *ASHRAE Journal* 2 (Sept. 1960): 62.

109. W. J. Stelpflug, "Effect of Modern Refrigeration on the Modern Supermarket," *Financial Analysts Journal* 10 (Nov. 1954): 63.

CHAPTER 6: "Who Ever Heard of an American without an Icebox?"

1. Before the invention of electric household refrigeration, people called iceboxes "refrigerators." We now call early non-electric refrigerators "iceboxes." So I am using anachronistic language in this chapter by referring to iceboxes as iceboxes instead of refrigerators. Quotations from primary sources, however, use the original word, "refrigerator," so you will want to remember this distinction for the rest of the chapter.

2. Blaisdell & Burley, *Blaisdell & Burley's Patent Elevating Refrigerator*, 1876, Western History Collection, Baker Library, Harvard Business School, Boston; Blaisdell & Burley, *Blaisdell's Elevating Refrigerator*, Tilton, NH, ca. 1875, Library Company of Philadelphia. I have looked high and low for additional literature related to Blaisdell & Burley and have only come up with these two flyers. To me, the paucity of surviving literature strongly suggests that Blaisdell & Burley's elevating refrigerator did not sell well, but I cannot prove it. Many thanks to the staff at the Library Company of Philadelphia for copying the 1875 flyer and sending it to me after I found it in their online catalog.

3. Rev. and Mrs. M. T. Runnels, quoted in *Blaisdell's Elevating Refrigerator*; Perkins Upright Refrigerator ad, quoted in Sarah F. MacMahon, "Laying Foods By: Gender, Dietary Decisions, and the Technology of Food Preservation in New England Households, 1750–1850," in *Early American Technology: Making & Doing Things from the Colonial Era to 1850*, ed. Judith A. McGaw (Chapel Hill: University of North Carolina Press, 1994), 178, n. 17.

4. Winifred James, quoted in Julie Greene, *The Canal Builders* (New York: Penguin, 2009), 71.

5. Oscar Edward Anderson, Jr., *Refrigeration in America* (Princeton: Princeton University Press, 1953), 114–15.

6. Barry Donaldson and Bernard Nagengast, *Heat & Cold: Mastering the Great Indoors* (Atlanta: American Society of Heating, Refrigerating and Air-Conditioning Engineers, 1994), 46.

7. Thomas Moore, *An Essay on the Most Eligible Construction of Ice-Houses: Also, a Description of the Newly Invented Machine Called the Refrigerator* (Baltimore: Bonsal & Niles, 1803), 18–19.

8. MacMahon, "Laying Foods By," 171–72.

9. Catharine Beecher, quoted ibid., 173.

10. Monica Ellis, *Ice and Icehouses through the Ages* (Southampton, UK: Southampton University Industrial Archeology Group, 1982), 20.

11. Washington, quoted in Paul Leland Haworth, *George Washington: Farmer* (Indianapolis: Bobbs-Merrill, 1915), 302.

12. Madison Cooper, *Practical Cold Storage*, 2nd ed. (Chicago: Nickerson Collins, 1914), 710.

13. Donaldson and Nagengast, *Heat & Cold*, 51.

14. Louise Jenison Peet and Lenore E. Sater, *Household Equipment* (New York: John Wiley & Sons, 1934), 174.

15. U.S. Patent Office, *List of Patents for Inventions and Designs* . . . (Washington, DC: J. and G. S. Gideon, 1847), 313.

16. *Cleveland Daily Herald*, 26 Feb. 1841.

17. Susan Strasser, *Never Done: A History of American Housework* (New York: Pantheon, 1982), 20.

18. Joseph C. Jones, Jr., *America's Icemen* (Olathe, KS: Jobeco Books, 1984), 140.

19. Joseph C. Jones, Jr., *American Ice Boxes* (Humble, TX: Jobeco Books, 1981), 46–47, 66.

20. Anne Francis Miksinski, "The Quiet Revolution: The History and Effects of Domestic Refrigeration in America; From Ice to Mechanical Refrigerants," M.A. thesis, George Washington University, 1970, 13.

21. *Scientific American* 11 (24 Nov. 1855): 82.

22. Ibid. (23 Aug. 1856): 394.

23. Donaldson and Nagengast, *Heat & Cold*, 51–52.

24. Jones, *America's Icemen*, 138.

25. Catharine E. Beecher, *Treatise on Domestic Economy*, rev. ed. (New York: Harper & Brothers, 1856), 322.

26. C. H. Leonard, *My Booklet on the Selection and Care of Refrigerators*, ca. 1890, 17.

27. Stock, quoted in Katheryn Krotzer Laborde, *Do Not Open: The Discarded Refrigerators of Post-Katrina New Orleans* (Jefferson, NC: McFarland, 2010), 117.

28. Elizabeth O. Hiller, *Left-Over Foods and How to Use Them*, McCray Refrigerator Company, 1910, 10, item 17, box 63, in Culinary Ephemera, William L. Clements Library, University of Michigan, Ann Arbor.

29. Northern Refrigerator Co., *Furniture City Refrigerators*, 1915–16, 26.

30. Sylvia Lovegren, *Fashionable Food: Seven Decades of Food Fads* (New York: Macmillan, 1995), 10.

31. Mae Savell Croy, *1000 Shorter Ways around the House* (New York: G. P. Putnam's Sons, 1916), 6.

32. G. M. Shirk Manufacturing Co., *Illustrated Catalogue and Price List of North Star Refrigerators*, Chicago, 1893, 5, item 1, box 67, in Culinary Ephemera, Clements Library.

33. Loretto Basil Duff, *A Course in Household Arts*, pt. 1 (Boston: Whitcomb and Barrows, 1916), 16–17.

34. Jones, *American Ice Boxes*, 53–54, 63.

35. *New York Tribune*, 28 April 1912.

36. Herrick Refrigerator Company, *Refrigerators for All Purposes*, Catalogue no. 5, ca. 1904, 7.

37. A. R. Norton, "Few Persons Know How to Make Wild or Tame Ice Do Its Duty," *New York Tribune*, 25 April 1915, 5.

38. Catherine Owen, *Progressive Housekeeping* (Boston: Houghton Mifflin, 1889), 105–6.

39. Boston Scientific Refrigerator Company catalog, 1877–78, 8.

40. *Philadelphia Inquirer*, 6 Aug. 1893, 22.

41. *Idaho Daily Statesman*, 2 Aug. 1903, 3.

42. Roger Thévenot, *A History of Refrigeration throughout the World*, trans. J. C. Fidler (Paris: International Institute of Refrigeration, 1979), 172.

43. *Manufacturer and Builder* 3 (March 1871): 56.

44. J. F. Buchanan & Co., *How Do We Keep Our Food?* ca. 1920, 3–4.

45. Norton, "Few Persons Know," 5.

46. Anderson, *Refrigeration in America*, 113.

47. Isobel Brands, "The 'Laws' of the Refrigerator," *Duluth News-Tribune*, 28 June 1916, 6.

48. McMahan, "Laying Foods By," 178, n. 17.

49. Catharine E. Beecher, "American People Starved," *Harper's New Monthly Magazine* 32 (May 1866), 766.

50. H. B. Hull, *Mechanical Refrigeration*, 4th ed. (Chicago: Nickerson & Collins, 1933), 80.

51. D. Eddy & Sons, *Eddy's Refrigerators*, Boston, 1894, 7.

52. This is especially true about putting warm things in the refrigerator, but more for reasons of bacterial growth than condensation.

53. J. C. Goosman, "History of Refrigeration," *Ice and Refrigeration* 67 (Aug. 1924): 111.

54. Thévenot, *History of Refrigeration throughout the World*, 172.

55. *White Cloud (KS) Chief*, 18 Aug. 1870, 1.

56. *Farmers' Cabinet*, 23 Aug. 1866, 5.

57. Ellis, *Ice and Icehouses*, 5.

58. *London Times*, May 1867 as quoted in Sylvia P. Beamon and Susan Roaf, *The Ice-Houses of Britain* (London: Routledge, 1989), 44.

59. *Ice and Refrigeration* 19 (Sept. 1900): 103; also *Milwaukee Sentinel*, 28 Sept. 1895, 7.

60. Syndicated columnist F. J. Haskins, quoted in *Ice* 1 (Aug. 1907): 24.

61. *Ice and Refrigeration* 25 (July 1903): 27.

62. U.S. Department of State, *Refrigerators and Food Preservation in Foreign Countries* (Washington, DC: GPO, 1891): 175–76, 173.

63. *Ice and Refrigeration* 21 (July 1901): 13.

64. Blanche McManus, "The Economies of the French Housewife," *American Cookery* 21 (April 1917): 672.

65. Department of State, *Refrigerators and Food Preservation in Foreign Countries*, 119, 127.

66. Ibid., 143, 49, 50.

67. Boston Scientific Refrigerator Company catalog, 1877–78, 6–7.

68. World's Columbian Exposition, *Report of the Committee on Awards* (Washington, DC: GPO, 1901), 2:1372.

69. John R. Williams, "A Study of Refrigeration in the Home and the Efficiency of Household Refrigerators," *Proceedings*, Third International Congress of Refrigeration (Chicago, 1913), 3:11.

70. *New York Tribune*, 9 May 1914, 7.

71. Williams, "A Study of Refrigeration in the Home," 13.

72. *New York Tribune*, 13 April 1919, F2.

73. Monroe Refrigerator Company, *The Monroe*, 1915. The year is provided by a letter accompanying the pamphlet in the Mohonk Mountain House Collection, Hagley Museum and Library, Wilmington, DE.

74. Maine Manufacturing Company, *White Mountain Refrigerators*, 1915, Nashua, NH, 3.

75. Donaldson and Nagengast, *Heat & Cold*, 52.

76. Robert Coit Chapin, *The Standard of Living among Workingmen's Families in New York City* (Philadelphia: William F. Fell, 1909), 136.

77. Thévenot, *History of Refrigeration throughout the World*, 172.

78. Shelley Nickles, "'Preserving Women': Refrigerator Design as Social Process in the 1930s," *Technology and Culture* 43 (Oct. 2002): 704.

79. Jones, *American Ice Boxes*, 44–45.

80. Thévenot, *History of Refrigeration throughout the World*, 334.

81. *Scientific American* 52 (6 June 1885): 363.

82. Ruth Schwartz Cowan, *More Work for Mother* (New York: Basic Books, 1983), 130–31.

83. *Scientific American* 99 (17 Oct. 1908): 257.

84. Thévenot, *History of Refrigeration throughout the World*, 173.

85. Mary Pattison, *Principles of Domestic Engineering* (New York: Trow Press, 1915), 94.

86. Donaldson and Nagengast, *Heat & Cold*, 208–9, 213.

87. Isko Co., *Refrigeration without Ice*, Detroit, ca. 1918, 2.

88. Donaldson and Nagengast, *Heat & Cold*, 215.

89. Alexander Stevenson, Jr., "Report on Domestic Refrigerating Machines, 1923–1925," transcribed by Anne Marie Nagengast, www.ashrae.org/aboutus/page/150, 400–402, accessed 5 Nov. 2010. Because of extra material that accompanies this document online, I use the pagination on the PDF rather than the pagination of the memo itself.

90. Ibid., 77.

CHAPTER 7: The Early Days of Electric Household Refrigeration

1. Frigidaire sales booklet, quoted in Joseph C. Jones, Jr., *American Ice boxes* (Humble, TX: Jobeco Books, 1981), 42–44. If there was a script for selling Frigidaires to men, I haven't seen it. However, this story from a Kelvinator sales booklet suggests how such a conversation might have gone:

> Sometimes a woman customer is interested in the mechanism, but more frequently it is the man. A successful salesman in a New Jersey city says that he always has to talk technical details to the husband, though rarely to the wife. "Generally, on my second call, when I have to see the husband and wife, I concentrate my attention mainly on the husband. I find that he wants to know the mechanism of the Kelvinator and how it works. He is accustomed to talking technical details of automobiles and of radio, and it is only natural that he should ask the same type of questions about electric refrigeration. I give him a good talk about Kelvinator mechanics. I keep it simple and free of technical language, but when I get done the prospect knows how Kelvinator works and he has a pretty good understanding of why it is the best type of electric refrigeration. Kelvinator Institute, *How Kelvinator Works*, 1925, 6–7.

2. Frigidaire Corporation, *Sales Training*, 1964, 3, box 2, folder 2, Frigidaire Historical Collection, Dunbar Library, Wright State University, Dayton, OH.

3. Kelvinator, *Kelvinator—and Some of Its Users*, ca. 1922, Refrigeration and Miscellaneous Pamphlets, National Museum of American History, Washington, DC.

4. Richard Rhodes, *The Making of the Atomic Bomb* (New York: Simon and Schuster, 1986), 20–21.

5. *Electrical World* 36 (July 1924): 28.

6. *The New Yorker*, quoted in Sylvia Lovegren, *Fashionable Food: Seven Decades of Food Fads* (New York: Macmillan, 1995), 10.

7. Roland Marchand, *Advertising the American Dream* (Berkeley: University of California Press, 1985), 269–73.

8. Alexander Stevenson, Jr., "Report on Domestic Refrigerating Machines, 1923–1925," transcribed by Anne Marie Nagengast, www.ashrae.org/aboutus/page/150, 19 (PDF pagination), accessed 5 Nov. 2010.

9. Ruth Schwartz Cowan, *More Work for Mother* (New York: Basic Books, 1983), 131. While one might assume that a refrigerator would just plug into household electric outlets,

that was not always the case. There were gas-powered refrigerators. Also, many Americans still did not have electric power during the 1920s.

10. Bernard Nagengast, "Electrical Refrigerators Vital Contribution to Households," *100 Years of Refrigeration*, A Supplement to *ASHRAE Journal* (Nov. 2004): S3.

11. Stevenson, "Report on Domestic Refrigerating Machines," 81.

12. Ibid., 82.

13. J. F. Nickerson, "The Household Machine as the Competitor of Ice," *Ice and Refrigeration* 37 (July 1925): 34.

14. Ronald C. Tobey, *Technology as Freedom: The New Deal and the Electrical Modernization of the American Home* (Berkeley: University of California Press, 1996), 19.

15. *Electrical Merchandising* 36 (Dec. 1926): 77.

16. Stevenson, "Report on Domestic Refrigerating Machines," 306, 409, 463.

17. *Scientific American* 133 (Sept. 1925): 194.

18. Frigidaire Corporation, *The Origin of Frigidaire*, 1964, 17, box A, folder 4, Frigidaire Historical Collection, Dunbar Library.

19. Frigidaire Corporation, *Frigidaire Service*, 1964, 1c, box 2, file 5, Frigidaire Historical Collection.

20. Nagengast, "Electrical Refrigerators Vital," S5.

21. Stevenson, "Report on Domestic Refrigerating Machines," 23–24.

22. *Ice and Refrigeration* 66 (May 1924): 460.

23. *Electrical World* 87 (15 May 1926): 1062.

24. New York Edison Company, *There Is More Leisure for the Housekeeper in Electrical Housekeeping*, 1924, 35, box 158, item 12, Culinary Ephemera, William L. Clements Library, University of Michigan, Ann Arbor.

25. Tobey, *Technology as Freedom*, 19, 18; Bernard Nagengast to Jonathan Rees (e-mail), 8 Dec. 2010.

26. Neil H. Borden, *The Economic Effects of Advertising* (Chicago: Richard D. Irwin, 1947), 398.

27. National Electric Light Association, *Proceedings* 81 (1924): 458.

28. Barry Donaldson and Bernard Nagengast, *Heat & Cold: Mastering the Great Indoors* (Atlanta: American Society of Heating, Refrigerating and Air-Conditioning Engineers, 1994), 220.

29. National Electric Light Association, *Proceedings* 83 (1926): 281.

30. David Nye, *Electrifying America: Social Meanings of a New Technology, 1880–1940* (New York: Cambridge University Press, 1990), 276.

31. *Ice and Refrigeration* 67 (Sept. 1924): 197.

32. *Ice and Refrigeration* 69 (Dec. 1925): 341.

33. Borden, *Economic Effects of Advertising*, 402–3.

34. National Electric Light Association, *Proceedings* 82 (1925): 511.

35. *Electrical Merchandising* 36 (Dec. 1926): 76, 78.

36. Copeland Sales Company, *How to Use Your Copeland*, ca. 1925, 18, box 60, item 3, Culinary Ephemera, Clements Library.

37. Borden, *Economic Effects of Advertising*, 396–97.

38. Anne Francis Miksinski, "The Quiet Revolution: The History and Effects of Domestic Refrigeration in America: From Ice to Mechanical Refrigerants," M.A. thesis, George Washington University, 1970, 90.

39. Tobey, *Technology as Freedom*, 19.

40. John E. Starr, “Refrigerating Fakes,” *Refrigerating World* 57 (Feb. 1922): 11–12.

41. Frigidaire Corporation, *Frigidaire Household Refrigerators*, 1964, 5, box 1A, file 23, Frigidaire Historical Collection, Dunbar Library.

42. Borden, *Economic Effects of Advertising*, 569.

43. Ibid., 397, 412.

44. Cowan, *More Work for Mother*, 139–43.

45. Miksinski, “Quiet Revolution,” 93–94, 109.

46. Susan Strasser, *Never Done: A History of American Housework* (New York: Pantheon, 1982), 265.

47. Frigidaire Corporation, *The Frigidaire Story*, 1966, 3; Miksinski, “Quiet Revolution,” 52–53; Donaldson and Nagengast, *Heat & Cold*, 211, 214; Frigidaire, *Frigidaire in Brief*, box 2, file 2, TOM (Tired Old Men) Club Collection, Dunbar Library, Wright State University.

48. Frigidaire, *Frigidaire Household Refrigerators*, 17, 23.

49. Oscar Edward Anderson, Jr., *Refrigeration in America* (Princeton: Princeton University Press, 1953), 195; Donaldson and Nagengast, *Heat & Cold*, 211.

50. *Refrigerating and Engineering News* 28 (Dec. 1934): 358.

51. Kelvinator, *Report by an Investigating Committee of Architects and Engineers*, 10 March 1923; Gail Cooper, *Air-Conditioning America: Engineers and the Controlled Environment, 1900–1960* (Baltimore: Johns Hopkins University Press, 1998), 122–23.

52. *Refrigerating and Engineering News* 28 (Dec. 1934): 372.

53. Roger Thévenot, *A History of Refrigeration throughout the World*, trans. J. C. Fidler (Paris: International Institute of Refrigeration, 1979), 172.

54. Stevenson, “Report on Domestic Refrigerating Machines,” 32–33.

55. Nagengast, “Electrical Refrigerators Vital,” S7.

56. *Electrical Merchandising* 39 (February 1928): 18.

57. General Electric, *Simplified Electric Refrigeration*, 2–3, 11, box 60, item 11, Culinary Ephemera, Clements Library.

58. Thomas F. O'Boyle, *At Any Cost: Jack Welch, General Electric, and the Pursuit of Profit* (New York: Knopf, 1998), 125.

59. Anderson, *Refrigeration in America*, 196.

60. General Electric, *Answering Your Questions about Electric Refrigeration*, 1929.

61. Cowan, *More Work for Mother*, 136; Nagengast, “Electrical Refrigerators Vital,” S7.

62. Borden, *Economic Effects of Advertising*, 402, 569. According to Borden, this advertising campaign worked: “It is impossible to determine what extent the increase in sales should be attributed to persuasive selling methods, but it is concluded that adoption of mechanical refrigeration by consumers would have been much slower had passive selling methods been employed” (ibid., 410).

63. Servel, quoted in Donaldson and Nagengast, *Heat & Cold*, 231.

64. *Frigidaire* 1 (ca. 1920): 3.

65. G. B. Denford to Dan Smiley, 29 Aug. 1921, box 13, folder R-1a, Lake Mohonk Mountain House Collection, Hagley Museum and Library, Wilmington, DE.

66. Kelvinator Corporation, *Kelvinator Electric Refrigeration for the Home*, 1924.

67. Frigidaire Corporation, *Frigidaire*, 1927, box 66, Item 4, Culinary Ephemera, Clements Library.

68. Isko Company, *Refrigeration without Ice*, Detroit, ca. 1918, 10.

69. Ralph K. Potter, “How It Works If a Lamp Socket Is Your Omnipresent Ice Man,” *New York Tribune*, 23 July 1922, 43.

70. General Electric, *Answering Your Questions about Electric Refrigeration*, 1929.

71. Frigidaire, quoted in Jones, *American Iceboxes*, 45.

72. Kelvinator Corporation, *Kelvinator Refrigerates without Ice*, 1921.

73. Kelvinator, *For the Hostess*, ca. 1925, 6, box 62, item 7, Culinary Ephemera, Clements Library.

74. Servel Corporation, *Servel Coldery*, 4, box 67, item 2, Culinary Ephemera, Clements Library.

75. Mary Bellis, "Making Ice Cubes: The History of Ice Cube Trays," http://inventors.about.com/od/istartinventions/a/IceCube.htm, accessed 29 July 2010.

76. Kelvinator, *Ice Cubes: A Symbol of the Modern Hostess*, ca. 1925, n.p.

77. Lisa Mae Robinson, "Safeguarded by Your Refrigerator: Mary Engle Pennington's Struggle with the National Association of Ice Industries," in *Rethinking Home Economics: Women and the History of the Profession*, ed. Sarah Stage and Virginia B. Vincenti (Ithaca: Cornell University Press, 1997): 255.

78. A. Hardgrave, "Development in Household Refrigeration," *Ice and Refrigeration* 66 (April 1924): 330.

79. John Nickerson, "Ice Finance," *Ice and Refrigeration* 71 (Dec. 1926): 437.

80. *Ice and Refrigeration* 70 (June 1926): 657.

81. Susanne Freidberg, *Fresh: A Perishable History* (Cambridge: Harvard University Press, 2009), 42–43.

82. Van Rensselaer H. Greene, "Manufactured Ice Industry Comes to Life," *Power* 67 (1928): 950, clipping in the Roy Eilers Collection, National Museum of American History, Washington, DC.

83. *Ice and Refrigeration* 101 (July 1941): 5.

84. Robinson, "Safeguarded by Your Refrigerator." 256.

85. *Ice and Refrigeration* 70 (March 1926): 283. See also Anderson, *Refrigeration in America*, 115, n. 45.

86. Stevenson, "Report on Domestic Refrigerating Machines," 53.

87. *Ice and Refrigeration* 77 (July 1929): 38.

88. Miksinski, "Quiet Revolution," 79–80, 102.

89. *Ice and Refrigeration* 77 (July 1929): 38.

90. Frigidaire Corporation, *The Story of Freon*, 1964, 1, box 1A, folder 3, Frigidaire Historical Collection, Dunbar Library.

91. "Household Automatic Refrigerating Machines," *The Traveler's Standard*, ca. 1927, 50–56, box 13, Lake Mohonk Mountain House Collection, Hagley Museum and Library.

92. Rhodes, *Making of the Atomic Bomb*, 20–21.

93. Du Pont, *Freon: Safe Refrigerants and Propellants*, 1950, 2.

94. W. D. Humphrey to F. Sparre, "Du Pont Refrigerants—A Comparative Study of Our Activities in This Field . . . ," 30 March 1931, 11, box 1036, Records of the E. I. Du Pont de Nemours & Company, series II, pt. 2, Hagley Museum and Library.

95. Alexander A. McCormack, *Cold Logic: Refrigeration Principles* (Chicago: Nickerson & Collins, 1949), 35.

96. Andrew D. Althouse and Carl H. Turnquist, *Modern Electric and Gas Refrigeration* (Chicago: Goodheart-Willcox, 1937), 140.

97. Miksinski, "Quiet Revolution," 100.

98. Research Laboratories, Roessler and Hasslacher Chemical Company, "Methyl Chloride—A Safe Refrigerant," Technical Paper 26, New York, 1926, 3.

99. Miksinski, "Quiet Revolution," 79.

100. Realto E. Cherne, "Developments in Refrigeration as Applied to Air Conditioning," *Ice and Refrigeration* 101 (July 1941): 28–29.

101. McCormack, *Cold Logic*, 36.

102. David A. Hounshell and John Kenly Smith, Jr., *Science and Corporate Strategy: Du Pont R&D, 1902–1980* (New York: Cambridge University Press, 1988), 156. Du Pont bought out GM's stake in Kinetic Chemicals in 1949 (Frigidaire, *The Story of Freon*, 9).

103. Frigidaire, *The Story of Freon*, 6.

104. Miksinski, "Quiet Revolution," 83–84.

105. Hounshell and Smith, *Science and Corporate Strategy*, 157.

106. Thévenot, *History of Refrigeration throughout the World*, 253.

CHAPTER 8: The Completion of the Modern Cold Chain

1. Kathleen Ann Smallzried, *The Everlasting Pleasure* (New York: Appleton-Century-Crofts, 1956), 246.

2. E. B. White, *Here Is New York* (New York: Harper, 1949), 28.

3. Roger Thévenot, *A History of Refrigeration throughout the World*, trans. J. C. Fidler (Paris: International Institute of Refrigeration, 1979), 334; Vivian Manufacturing Company, *Ice Tools and Supplies* (catalog), St. Louis, MO, 1949, 39.

4. U.S. Department of Commerce, Bureau of the Census, *Biennial Census of Manufactures, 1935* (Washington, DC: GPO, 1938), 8:157.

5. Anne Francis Miksinski, "The Quiet Revolution: The History and Effects of Domestic Refrigeration in America; From Ice to Mechanical Refrigerants," M.A. thesis, George Washington University, 1970, 106.

6. Thévenot, *History of Refrigeration throughout the World*, 173.

7. Neil H. Borden, *The Economic Effects of Advertising* (Chicago: Richard D. Irwin, 1947), 413.

8. W. Michael Cox and Richard Alm, "You Are What You Spend," *New York Times*, 10 Feb. 2008.

9. Ronald C. Tobey, *Technology as Freedom: The New Deal and the Electrical Modernization of the American Home* (Berkeley: University of California Press, 1996), 165. The owners of large apartment buildings often bought refrigerators for their units, but Tobey's work does not suggest that such arrangements existed in Riverside.

10. Ibid., 165, 19.

11. Borden, *Economic Effects of Advertising*, 411.

12. Tobey, *Technology as Freedom*, 139, 162.

13. Shelley Nickles, "'Preserving Women': Refrigerator Design as Social Process in the 1930s," *Technology and Culture* 43 (Oct. 2002): 724–26.

14. Tobey, *Technology as Freedom*, 175.

15. *House and Garden*, quoted in Miksinski, "Quiet Revolution," 113.

16. Lawrence C. Lockley, "The Turn-over of the Refrigerator Market," *Journal of Marketing* 2 (Jan. 1938): 210.

17. Tobey, *Technology as Freedom*, 123.

18. Bernard Nagengast, "Electrical Refrigerators Vital Contribution to Households," *100 Years of Refrigeration*, A Supplement to *ASHRAE Journal* (Nov. 2004): S8.

19. Borden, *Economic Effects of Advertising*, 573.

20. Oscar Edward Anderson, Jr., *Refrigeration in America* (Princeton: Princeton University Press, 1953), 220–21; Katheryn Krotzer Laborde, *Do Not Open: The Discarded Refrigerators of Post–Katrina New Orleans* (Jefferson, NC: McFarland, 2010), 117.

21. Philip Roth, *Goodbye, Columbus* (1959; New York: Vintage, 1993), 42–43.

22. David Owen, "The Efficiency Dilemma," *New Yorker* 86 (20 and 27 Dec. 2010): 78.

23. Borden, *Economic Effects of Advertising*, 570.

24. General Electric, *1939: The Beautiful New General Electric Triple-Thrift Refrigerators*, 14.

25. Frigidaire Corporation, *Frigidaire Service*, 37, box 2, folder 5, Frigidaire Historical Collection, Dunbar Library, Wright State University, Dayton, OH.

26. *Changing Times* 43 (Jan. 1989): 84.

27. *Financial Times*, 20 Feb. 1956, 21.

28. Nickles, "Preserving Women," 720.

29. John Holusha, "The Refrigerator of the Future, for Better and Worse," *New York Times*, 30 Aug. 1992.

30. *Appliance* 28 (Jan. 1971): 26.

31. Miksinski, "Quiet Revolution," 150–51.

32. Morris Kaplan, "Quality Is What the Consumer Wants," *ASHRAE Journal* 3 (Nov. 1961): 81.

33. Sandy Isenstadt, "Visions of Plenty: Refrigerators in America around 1950," *Journal of Design History* 11 (1998): 311.

34. *Good Housekeeping* 130 (June 1950): 26.

35. Louise Jenison Peet and Lenore E. Sater, *Household Equipment* (New York: John Wiley & Sons, 1934), 188.

36. Borden, *Economic Effects of Advertising*, 572.

37. Organisation for European Economic Co-operation, *The Cold Chain in the U.S.A.*, pt. 2 (Paris: OEEC, 1952), 115.

38. Frigidaire, *A Special Offer for You . . . A Dayton Union Member . . .*, ca. 1972, 4, 6, box 8, folder 5, Frigidaire Historical Collection, Dunbar Library.

39. Lowe's, *Refrigerator Buyer's Guide*, 2007, 3.

40. Frigidaire Corporation, *Sales Training*, 1964, 53, box 2, folder 2, Frigidaire Historical Collection.

41. *Life* 19 (5 Nov. 1945): 28.

42. Helen Kendell and George A. Papritz, "Shopping for Your New Refrigerator," *Good Housekeeping* 129 (July 1949): 132.

43. *Life* 56 (28 Feb. 1964): 85.

44. Jonathan Bloom, *American Wasteland: How America Throws Away Nearly Half Its Food (and What We Can Do about It)* (Cambridge, MA: De Capo Press, 2010), 196.

45. Kate Ramsayer, "Do Your Homework Before Buying a New Eco-Friendly Refrigerator," *Bend (OR) Bulletin*, 22 Feb. 2010.

46. Ben-Hur Manufacturing Co., *What's in This Modern Ben-Hur*, 1952, Box 1, [93–8466], Refrigeration, Warshaw Collection, National Museum of American History, Washington, DC.

47. National Electrical Manufacturers Association, *How to Enjoy Better Meals with Less Work at Lower Cost*, 3, 9, box 64, item 23, Culinary Ephemera, William L. Clements Library, University of Michigan, Ann Arbor.

48. Harry Carlton, *The Frozen Food Industry* (Knoxville: University of Tennessee Press, 1941), 2.

49. Frigidaire, *The New Thrills of Freezing with Your Frigidaire Food Freezer*, 1, box 64, item 6, Culinary Ephemera, Clements Library.

50. Shirley Rolfs, *Freezing Foods at Home* (Minneapolis: Van Wold Stevens, 1946), 3.

51. Mark Kurlansky, *Birdseye: The Adventures of a Curious Man* (New York: Doubleday, 2012), 137–39.

52. Clarence Francis, *A History of Food and Its Preservation* (Princeton: Princeton University Press, 1937), 12–13.

53. Kurlansky, *Birdseye*, 157–59.

54. Andrew F. Smith, *Eating History: 30 Turning Points in the Making of American Cuisine* (New York: Columbia University Press, 2009), 167–68.

55. Carlton, *Frozen Food Industry*, 4–7.

56. Warren Belasco, *Meals to Come: A History of the Future of Food* (Berkeley: University of California Press, 2006), 175.

57. Kurlansky, *Birdseye*, 183 (including *New York Times* quotation).

58. Anderson, *Refrigeration in America*, 276–77.

59. OEEC, *Cold Chain in the U.S.A.*, pt. 1, 47, 49, and pt. 2, 279; Shane Hamilton, *Trucking Country: The Road to America's Wal-Mart Economy* (Princeton: Princeton University Press, 2008), 119–20.

60. Francis, *History of Food*, 32–33.

61. *Appliance* 28 (Jan. 1971): 33.

62. H. B. Hull, *Mechanical Refrigeration*, 4th ed. (Chicago: Nickerson & Collins, 1933), 57.

63. *Ice and Refrigeration* 101 (Aug. 1941): 217–18.

64. Ibid. (July 1941): 49–53.

65. Albert Todoroff, *How to Build and Operate a Locker Plant* (St. Louis: Meat Merchandising, 1944), 11–12.

66. National Frozen Food Locker Association, *Frozen Food Suggestions*, 1947, 8, box 68b, item 15, Culinary Ephemera, Clements Library.

67. Nagengast, "Electrical Refrigerators Vital," S8.

68. Bloom, *American Wasteland*, 189.

69. Frigidaire Corporation, *Frigidaire and Frozen Foods*, 2, box 1ab, folder 4, Frigidaire Historical Collection, Dunbar Library.

70. Donald K. Tressler, Clifford F. Evers, and Barbara Hutchings Evers, *Into the Freezer—and Out*, 2nd ed. (New York: Avi, 1953), 11.

71. Ibid., 7.

72. Internet Movie Database, *I Love Lucy: The Freezer*, www.imdb.com/title/tt0609355, accessed 2 Jan. 2011. IMDB summarizes the entire episode as follows: "Lucy and Ethel decide to make some money by buying a used walk-in freezer. When they unwittingly buy 700 pounds of beef, they try to sell it to customers at the butcher shop."

73. American Motors Corporation, *Your New Home Freezer*, 1959, 10, box 64, item 8, Culinary Ephemera, Clements Library.

74. Good Reading Rack Service, *Freezer Handbook*, 1963, 4, box 68b, item 27, Culinary Ephemera.

75. Sears, *Freezer Living Is Good Living with Coldspot*, ca. 1968, 7, box 68b, item 28, Culinary Ephemera.

76. Miksinski, "Quiet Revolution," 142–43.

77. Robert L. Olson, "Prepared Frozen Foods," *ASHRAE Journal* 8 (July 1966): 56–57.

78. Tracie McMillan, *The American Way of Eating: Undercover at Walmart, Applebee's, Farm Fields and the Dinner Table* (New York: Scribner, 2012), 185–232.

79. Alice Bradley, *Electric Refrigerator Menus and Recipes*, General Electric, 1927, 33.

80. Nagengast, "Electrical Refrigerators Vital," S8-S9.

81. Examples of the ads for such devices appear in, for example, the May and June 1954 issues of *Popular Mechanics* (pp. 33 and 25, respectively).

82. Hotpoint, *Your Hotpoint Combination Refrigerator-Freezer Instruction and Recipe Book*, 11, box 60, item 25, Culinary Ephemera, Clements Library.

83. Westinghouse, *1953 Westinghouse Refrigerators*, 4, box 67, item 16, Culinary Ephemera.

84. Frigidaire, *The Engineering Department*, 23, box 2, folder 3, Frigidaire Historical Collection, Dunbar Library.

85. Frigidaire, *Sales Training*, 35, box 2, folder 2, Frigidaire Historical Collection.

86. Ruth Schwartz Cowan, *More Work for Mother* (New York: Basic Books, 1983), 139.

87. *Times (London)*, 10 Dec. 1957, 8.

88. *Ibid.*, 6 March 1965, 11.

89. Hans-Liudger Dienel, *Linde: History of a Technology Corporation, 1879–2004* (New York: Palgrave Macmillan, 2004), 149.

90. Thévenot, *History of Refrigeration throughout the World*, 174, 348.

91. OEEC, *Cold Chain in the U.S.A.*, pt. 1, 32, 28.

92. U.S. Department of Commerce, *Major Household Appliances: Select Foreign Countries* (Washington, DC: GPO, 1960), 24.

93. Walter Pichel, "Quick Frozen Food in France," *ASHRAE Journal* 5 (May 1963): 77–78.

94. Department of Commerce, *Major Household Appliances*, 3.

95. Owen, "Efficiency Dilemma," 79.

96. Department of Commerce, *Major Household Appliances*, 109.

97. Walter Pichel, "Future Development of Household Refrigerators in Europe," *ASHRAE Journal* 6 (July 1964): 76–78.

98. Bill McKibben, "Can China Go Green?" *National Geographic* 219 (June 2011): 119.

99. Peter Menzel and Faith D'Aluisio, *Hungry Planet: What the World Eats* (Berkeley: Material World Books / Ten Speed Press, 2005), 84.

100. *The Economist*, 4 Aug. 2012, www.economist.com/node/21559977/print, accessed 8 Aug. 2012.

101. Yoon Ja-young, "Kimchi Refrigerator Maintains Taste of Fermented Food," *Korea Times*, 21 Nov. 2011, www.koreatimes.co.kr/www/news/tech/2012/03/133_99236.html, accessed 30 June 2012.

CONCLUSION: Refrigeration, Capitalism, and the Environment

1. Andrei Codrescu and Nils Juul-Hansen, "If These Refrigerators Could Speak," *New York Times*, 29 Jan. 2006. See also Katheryn Krotzer Laborde, *Do Not Open: The Discarded Refrigerators of Post-Katrina New Orleans* (Jefferson, NC: McFarland, 2010).

2. Codrescu and Hansen, "If These Refrigerators Could Speak."

3. Bruce Hackett and Loren Lutzenhiser, "The Unity of Self and Object," *Western Folklore* 44 (Oct. 1985): 320, 323.

4. Laborde, *Do Not Open*, 121–22.

5. Adam Gopnik, *The Table Comes First: Family, France, and the Meaning of Food* (New York: Knopf, 2011), 95.

6. *Shenandoah Herald*, 17 Sept. 1879, 1.

7. Twain, quoted in Andrew Beahrs, *Twain's Feast: Searching for America's Lost Foods in the Footsteps of Samuel Clemens* (New York: Penguin, 2010), 96.

8. Mark Twain, *A Tramp Abroad*, 3rd ed. (London: Chatto & Windus, 1880), 1:285.

9. Georg Johann Kohl, *Russia and the Russians*, quoted in Elizabeth David, *Harvest of the Cold Months* (New York: Viking, 1994), 295.

10. Walter Pichel, "Development of Refrigeration in USSR—Now and in the Next Five Years," *ASHRAE Journal* 7 (April 1965): 90.

11. Jenny Leigh Smith, "Empire of Ice Cream: How Life Became Sweeter in the Postwar Soviet Union," in *Food Chains: From Farmyard to Shopping Cart*, ed. Warren Belasco and Roger Horowitz (Philadelphia: University of Pennsylvania Press, 2009), 150.

12. T. R. Malthus, *An Essay on the Principle of Population* (London: T. Bensley, 1806), 2:71.

13. Werner Sombart, *Why Is There No Socialism in the United States?* trans. Patricia M. Hocking and C. T. Husbands (White Plains, NY: International Arts and Sciences Press, 1976), 106.

14. Adam Smith, *An Inquiry into the Nature and Causes of the Wealth of Nations*, bk. 1 (Chicago University of Chicago Press, 1976), 183.

15. Pierre Desrochers and Hiroko Shimzu, *The Locavore's Dilemma: In Praise of the 10,000 Mile Diet* (New York: Public Affairs, 2012), 11, 113–14.

16. The best available review of Chandler's career is Thomas K. McCraw, "Introduction: The Intellectual Odyssey of Alfred D. Chandler, Jr.," in *The Essential Alfred Chandler: Essays toward a Historical Theory of Big Business* (Cambridge: Harvard Business School Press, 1988), 1–21. Chandler's two most famous works are *The Visible Hand: The Managerial Revolution in American Business* (Cambridge: Belknap, 1977) and *Scale and Scope: The Dynamics of Industrial Capitalism* (Cambridge: Belknap, 1990).

17. Philip Scranton, *Endless Novelty: Specialty Production and American Industrialization, 1865–1925* (Princeton: Princeton University Press, 1997), 10–11.

18. Belasco and Horowitz, *Food Chains*.

19. William Cronon, *Nature's Metropolis: Chicago and the Great West* (New York: Norton, 1991), 56.

20. Colin Beavan, "Update on No Impact Land," No Impact Man, 7 June 2007, http://noimpactman.typepad.com/blog/2007/06/update_on_no_im.html, accessed 4 Aug. 2010.

21. Michelle Beavan, in *No Impact Man* (Oscilloscope Pictures, 2009), a documentary film made of the Beavans' project. Colin Beavan also recounted his and his family's experiences in a book, *No Impact Man* (New York: Farrar Strauss and Giroux, 2009).

22. Steven L. Hopp, "Oily Food," in Barbara Kingsolver, *Animal, Vegetable, Miracle: A Year of Food Life* (New York: Harper Collins, 2007), 5.

23. Organisation for Economic Co-operation and Development, *Energy Efficiency Standards for Traded Products*, Annex 1, Working Paper no. 5 (Paris: OECD 1998), 8.

24. James E. McWilliams offers an extensive discussion of the origins of this often-cited statistic as well as the many problems with the study that generated that result in *Just Food: Where Locavores Get It Wrong and How We Can Truly Eat Responsibly* (New York: Little, Brown, 2009), 19–21.

25. Mike Berners-Lee, *How Bad Are Bananas? The Carbon Footprint of Everything* (London: Profile Books, 2010), 27.

26. Stephen Budiansky, "Math Lessons for Locavores," *New York Times*, 19 Aug. 2010.

27. Anna Lappé, *Diet for a Hot Planet: The Climate Crisis at the End of Your Fork and What You Can Do about It* (New York: Bloomsbury, 2010), 11, 32.

28. Tom Shachtman, *Absolute Zero and the Conquest of Cold* (Boston: Houghton Mifflin, 1999), 16–17.

29. F. E. Matthews, "Improved Cold Storage Methods: A Means to Better World Provisioning," *A.S.R.E. Journal* 5 (May 1919): 417.

30. Andrew Martin, "One Country's Table Scraps, Another Country's Meal," *New York Times*, 18 May 2008. Because the amount of food produced in this country has shot up thanks to improved production techniques, the total amount of food wasted is much larger now than it was in 1917, despite the fact that the waste is a lower portion of the total produced. This only underscores the need to use refrigeration more effectively.

31. Lee A. Craig, Barry K. Goodwin, and Thomas Grennes, "The Effect of Mechanical Refrigeration on Nutrition in the United States," *Social Science History* 28 (Summer 2004): 327, 325.

32. Tristram Stuart, *Waste: Uncovering the Global Food Scandal* (New York: Norton, 2009), 233–34.

33. Anthony Bourdain, *A Cook's Tour: Global Adventures in Extreme Cuisine* (New York: Bloomsbury, 2001), 210–11.

34. Desrochers and Shimzu, *Locavore's Dilemma*, 63, xxi.

35. Greenpeace, *HFCs: A Growing Threat to the Climate*, Updated Edition (Amsterdam: Greenpeace, Dec. 2009), 10.

36. Ibid., 10.

37. AFP, "UN Scientists Say Ozone Layer Depletion Has Stopped," 16 Sept. 2010, www.google.com/hostednews/afp/article/ALeqM5go2RnKHmYcJjFtzH-vKYV2gs_yJA, accessed 1 Oct. 2010.

38. Elizabeth Kolbert, "Note to Detroit: Consider the Refrigerator," The New Yorker Blog, 11 Dec. 2008, www.newyorker.com/online/blogs/tny/2008/12/note-to-detroit-consider-the-r.html#ixzz2oyYTQ922, accessed 18 July 2012.

39. See Greenpeace, *HFCs*, 10. HFCs were hasty and temporary replacements for the fluorine gases during the 1990s when CFCs and HCFCs (both generic terms for gases like Freon) were found to deplete the ozone layer. While the newer of these refrigerants do not affect the ozone layer when they are released into the atmosphere, they do cause global warming when they escape, much more so than carbon dioxide, the gas created when products like coal are burned to provide the power that makes refrigerators run.

40. Ibid., 16.

Essay on Sources

During graduate school at the University of Wisconsin–Madison, I spent a lot of time on the fourth floor of the Engineering Library while working on my dissertation. My topic concerned the American steel industry, so I spent a lot of time reading hundred-year-old copies of the trade journal *Iron Age*. One day, coming up out of the elevator, I spotted a different trade journal and started thumbing through it. Its name was *Ice and Refrigeration*, and while glancing through it, I quickly learned that the United States had once had a gigantic natural ice industry. I also learned that refrigeration at the turn of the twentieth century was nothing like refrigeration today.

This book is built from trade journals like that one. *Ice and Refrigeration*, founded in 1891, was the bible of the ice and refrigeration industries in America and the world during its early years. I have also read through many volumes of other industry journals: *Ice and Cold Storage*, *Ice*, *Cold*, *Ice Trade Journal*. Since the electrical industry dabbled in mechanical refrigeration too, I have read many articles in that industry's trade journals as well.

As most trade journal articles are unsigned and short, I have cited them with just the name of the journal, date, volume and page numbers. All articles credited with an author are cited with the author's name and the title of the article. I first found many of the articles I cite (particularly the ones from electrical industry journals) in the Roy Eilers Collection at the National Museum of American History in Washington, D.C., during a month-long fellowship there in 2000. Eilers was a St. Louis patent attorney who kept his own clipping service of technologies of all kinds. By a stroke of luck, the first set of clippings processed by the archivists there turned out to be on the refrigeration industry. Unfortunately, Eilers did not stamp volume numbers on the clips that he cut out, just journal, date, and page. In order to improve the accessibility of my sources to future researchers, I have found as many of those clips as I could in the original journal and cited them in full here. If I could not find the volume number, I have included a reference to the Eilers Collection in my citation.

The second most important group of sources for this book were the catalogs and literature of ice and refrigeration companies. These are all published materials, but again often extremely rare. Most libraries, like the Library of Congress and the National Museum of American History, file these pamphlets as books and keep them in the stacks. Others, like the Hagley Museum and Library in Delaware, have some of this material in the stacks and some in archival collections. My rule has been that if I found the catalog in a collection, I have cited it as I would any other archival source. If I found it in the stacks of a library, I have cited it as I would a published source.

As for published books, there is a notable absence of historical work on the American ice and refrigeration industries. As a result, I owe a great debt to four works that do consider this subject. Oscar Edward Anderson's *Refrigeration in America: A History of a New Technology and*

Its Impact (Princeton: Princeton University Press, 1953) is the last such work by an academic historian. It is really more of a traditional business history than a serious examination of the technology of ice or refrigeration. Nevertheless, it was still very helpful. Roger Thévenot's *History of Refrigeration throughout the World* (trans. J. C. Fidler; Paris: International Institute of Refrigeration, 1979) and Barry Donaldson and Bernard Nagengast's *Heat & Cold: Mastering the Great Indoors* (Atlanta: American Society of Heating, Refrigeration and Air Conditioning Engineers, 1994) are both the work of engineers. All three of these books are out of print and exceedingly rare. The geographer Susanne Freidberg's *Fresh: A Perishable History* (Cambridge: Harvard University Press, 2009) includes a single chapter on ice and refrigeration and then concentrates on how refrigeration affected the consumption of different types of food. It is almost exclusively consumer-oriented, but Freidberg's argument that freshness is culturally dependent still had a great impact upon my thinking.

Newspapers have been another important source for this book. In this instance, I have benefited greatly from the new technology of electronic databases. Ice and refrigeration came up often in the press so often around the turn of the twentieth century that there is a wealth of information on it out there. However, if you look up "ice" in a database like Chronicling America from the Library of Congress (http://chroniclingamerica.loc.gov), you will be drowning in material. I used what I learned from trade journals to guide my search in multiple databases, Chronicling America and Nineteenth Century American Newspapers being the two most important. All the articles on the fire at the Columbian Exposition I found the old-fashioned way (on microfilm) at the Chicago Public Library.

There are not that many archives with collections devoted to ice and refrigeration, but I did use a few. The Frederic Tudor Papers at the Harvard Business School and the Frigidaire Historical Collection at Wright State University were the two most important. Others, like the Mohonk Mountain House Papers at the Hagley Museum and Library were good for finding catalogs, but I have cited the catalogs from those collections only if I could not find them elsewhere. The National Museum of American History is also better for catalogs than for archival material, but I found some interesting resources in the Warshaw Collection.

I started this project long before Google began scanning books. Thanks to those efforts, many extremely rare pre-1923 books and journals are both easily accessible and searchable. Google Books is an absolute godsend to historians working in the Gilded Age and Progressive Era, as so many otherwise hard-to-find resources are now easily available. Many of the rarest ice and refrigeration journals, however, remain unscanned at this writing, so I read them on microfilm or in gigantic repositories such as the New York Public Library or the Library of Congress. I am grateful to have gotten my initial introduction to these industries mostly through good old-fashioned paper, as it provided me with a strong foundation for future electronic searching.

Index